11/29/21

STRATEGIYA

OFER FRIDMAN
(*Translator and Editor*)

Strategiya

The Foundations of the Russian Art of Strategy

OXFORD
UNIVERSITY PRESS

Oxford University Press is a department of the
University of Oxford. It furthers the University's objective
of excellence in research, scholarship, and education
by publishing worldwide.

Oxford   New York
Auckland   Cape Town   Dar es Salaam   Hong Kong   Karachi
Kuala Lumpur   Madrid   Melbourne   Mexico City   Nairobi
New Delhi   Shanghai   Taipei   Toronto

With offices in
Argentina   Austria   Brazil   Chile   Czech Republic   France   Greece
Guatemala   Hungary   Italy   Japan   Poland   Portugal   Singapore
South Korea   Switzerland   Thailand   Turkey   Ukraine   Vietnam

Oxford is a registered trade mark of Oxford University Press
in the UK and certain other countries.

Published in the United States of America by
Oxford University Press
198 Madison Avenue, New York, NY 10016

Copyright © Ofer Fridman 2021

All rights reserved. No part of this publication may be reproduced,
stored in a retrieval system, or transmitted, in any form or by any means,
without the prior permission in writing of Oxford University Press,
or as expressly permitted by law, by license, or under terms agreed with
the appropriate reproduction rights organization. Inquiries concerning
reproduction outside the scope of the above should be sent to the
Rights Department, Oxford University Press, at the address above.

You must not circulate this work in any other form
and you must impose this same condition on any acquirer.

Library of Congress Cataloging-in-Publication Data is available
Ofer Fridman.
Strategiya: The Foundations of the Russian Art of Strategy.
ISBN: 9780197606162

Printed in Great Britain by Bell and Bain Ltd, Glasgow

The national aspects of our military art show that the Russian people harbour an enormous reserve of creativity in the field of military art, that our army has wealth and strength of thought and spirit, and that it does not need any tutorage in this respect, no matter where it comes from.

Aleksei Baiov (1913)[1]

A body of contemporary political and sociological knowledge on the questions of military strategy would be flawed without referring to our forgotten or half-forgotten native authors, who conducted fruitful work before the 1917 Revolution and in certain periods of Soviet history.

Andrey Kokoshin (2016)[2]

[1] 'Natsional'nyye cherty russkogo voyennogo iskusstva v romanovskiy period nashey istorii' [National Features of Russian Military Art in the Romanov Period of Our History] (author's translation).
[2] *Politologiya i sotsiologia voennoi strategii* [The Political Science and Sociology of Military Strategy], (Moscow: URSS), p. 34 (author's translation).

CONTENTS

Acknowledgments	ix
Foreword by Professor Sir Lawrence Freedman	xi
Editor's Introduction	1
1. The Experience of Historical-Critical Research into the Laws of Military Art (Positive Strategy)	
Genrikh Antonovich Leer	23
2. The Responsibilities of Politics in Its Relations with Strategy	
Evgeny Ivanovich Martynov	75
3. The Foundations of Strategy	
Nikolai Petrovich Mikhnevich	107
4. The Philosophy of War	
Anton Antonovich Kersnovski	137
5. The Science of War: On Sociological Research into War	
Nikolai Nikolayevich Golovin	199
6. The Face of Contemporary War	
Evgeny Eduardovich Messner	235
About the Authors	289
Index of Names	295
Index of Places, Wars and Battles	303

ACKNOWLEDGEMENTS

This project would not have been possible without the generous and invaluable assistance of many people, to whom I would like to express my gratitude. Special thanks are due to Beatrice Heuser, to whom I owe a great deal. Though her responsibilities as my PhD supervisor ended long ago, her continuous support of all my endeavours has been absolutely invaluable. I also wish to express my special appreciation to Neville Bolt, whose readiness to help and unlimited support went far beyond my expectations.

I would like to thank all members of the Department of War Studies at King's College, with whom I happened to discuss this project. Their support, as well as criticism, served as the fuel that fed the engine of this project.

Surprisingly (or not), I feel indebted to the many scholars whose works I have been translating for the last year and a half. I am sure most of them would be very surprised (and probably pleased) to know that at the beginning of the third decade of the 21st century, their works have been translated into English. Trying to get into their minds through their work was a very interesting and intellectually challenging journey, and for that I am grateful to them.

I would also like to express my gratitude to friends and relatives, of whom there are too many to list, who probably never quite understood what I was doing or why, but always believed in me and in my project nevertheless. I thank you all for believing in me even when I did not.

ACKNOWLEDGEMENTS

Last, but definitely not least, I would like to thank the Gerda Henkel Foundation for supporting this project, and Thomas Podranski, manager of the special programme 'Security, Society and the State', who helped me through many intricate administrative issues.

FOREWORD

Major wars are rare events. When they occur, however, the costs and consequences are likely to be severe, even if they end with victory. This puts pressure on military strategists to anticipate and prepare for future conflicts which may never happen, though if they do the potential effects are likely to be devastating and long-lasting. To help them prepare they will carefully examine any recent war, however limited, for potential 'lessons' for the next, seeking to assess the impact of new technologies or the success of innovative tactics. But that is not enough. There can be no certainty that the next war will replicate the last, at least not sufficiently to allow any lessons to be applied with confidence. To broaden their perspectives, therefore, strategists have little choice but to look to history. That is why the military profession, possibly more than any other, is historically-minded. It studies it own history, not just out of intellectual curiosity, but to find sources of inspiration and guidance for the future.

This was a course advocated by Napoleon, who was wary of theoretical treatises on war. His advice for aspiring generals:

> 'Peruse again and again the campaigns of Alexander, Hannibal, Caesar, Gustavus Adolphus, Turenne, Eugene, and Frederick. Model yourself upon them. This is the only means of becoming a great captain, and of acquiring the secret of the art of war. Your own genius will be enlightened and improved by this study, and you will learn to reject all maxims foreign to the principles of the great commanders.'

Those who built their reputations as strategic thinkers, as interpreters of the Napoleonic Wars, such as Antoine-Henri, Baron de Jomini, turned

FOREWORD

this advice into dogma by insisting that the core principles of strategy were timeless. Through the intensive study of these great figures and their most famous battles, the secrets of success in war might be discerned. It was true that strategic practice would be shaped by the quality of the weaponry, broader social changes and the circumstances of any particular conflict, but they could always be viewed with the benefit of the established framework. So in addition to studying the great captains there have also been good reasons to study the great strategists.

In most countries there is a stream of strategic thinking that takes in their history and culture, focusing on particular conflicts, campaigns and battles, while addressing the ideas of key theorists and practitioners. In Russia this stream was broken by the combination of the trauma of the First World War and then the 1917 revolution. The influences of the past never went away but the new orthodoxies of Marxism-Leninism, and the need to proclaim the superior strategic wisdom of the Communist Party, provided the dominant narratives. After some seven decades, and another traumatic war, there was a need and an opportunity for independent strategic thinking, but the circumstances were very different.

This is why Ofer Fridman's translations of some important but hitherto neglected works of Russian strategists are so valuable. This is a contribution to scholarship on the development of Russian strategic thinking over the nineteenth and twentieth centuries. The works are fascinating examples of strategic thought in their own right. The texts from pre-revolutionary Russia show how much its strategic thought was tuned into the European mainstream, albeit with some distinctive Russian characteristics. This tradition did not die out after the Bolshevik Revolution, not least because some of the practitioners continued to serve in the Soviet Army. Eventually, however, it could only survive in exile. It is from these exiles that Fridman draws his second selection of readings.

These texts differ in style and content from the dreary publications that flourished in the Soviet Union, especially once the officer class had come to live in fear of Stalin. The stress on the scientific correctness of Marxism-Leninism, combined with centralised control of all aspects of military affairs, resulted in standard formulations that allowed little scope for individual interpretations. During the Cold War the main

FOREWORD

text was the Marshal Sokolovsky edited work *Soviet Military Strategy*, first published in 1962, produced by a collective of authors. This went through a number of editions. Western interest at each point was largely in the shifts from one edition to the next.

That can be contrasted with the exuberant prose of Genrikh Antonovich Leer, Fridman's first contributor. He insisted that strategy was an art, albeit one that was unusually difficult and complex. Moreover it was one that could not wait until inspiration struck and then developed in tranquillity. Once a war begins, inspiration must be found at every moment and in nerve-racking situations. And while it might be based on unchanging laws the application could vary infinitely because of constantly changing environment. For this reason it was situation-dependent and could not flow from excessive and uncritical dependence on heavy-handed theorising. Reading Leer it is hard not to be struck by the detailed references to a wide range of European theorists and practitioners, and the evident influences of figures such as Jomini and to a lesser extent Clausewitz. His approach to strategy as a 'synthesis of military affairs' is distinct from the prevailing, narrower definition of the time, which saw strategy as being largely about getting forces into a position to fight battles, with tactics taking over as the actual fighting began.

As the nineteenth century turned into the twentieth, the idea that the political and military spheres represented two distinct competences was increasingly untenable, for the military depended on the politicians for their budgets and bases, their alliances and their enemies. Yet the interaction between the two spheres was not seen as central to strategy. In these readings that is not the case. Politics is important not just in providing the objectives for which the military must fight, and setting the conditions which enable them to do so, but in providing a core idea about the nation and the state which inspires armies and shapes the way they fight. This encourages a nationalistic celebration of a strategic culture. Thus, in 1899, Evgeny Ivanovich Martynov concludes by praising not only the close relationship between Russia's military and diplomatic departments but also the autocracy that allows both to act with 'secrecy and decisiveness'.

Despite the political insights that inform Russia's pre-World War One strategic thought, when the war came diplomacy struggled and

FOREWORD

the military performance was inadequate. A sense of regret and grievance runs through Anton Antonovich Kersnovski's *Philosophy of War*, written on the eve of the next great conflagration. He combines strategic insights and historical commentary with a conviction that Russian strategy must reflect its national, Christian, spiritual roots and avoid the soulless rationalism of Lenin. By the time we reach Evgeny Eduardovich Messner (who had worked with the Wehrmacht during the Second World War), writing in 1959, the alienation from the main trends of Western strategic thought is almost complete, and instead there is a sardonic commentary on geopolitical trends, the impact of thermonuclear weapons and the role of propaganda.

The contemporary strategist may be both stimulated and shocked by this unique set of writings, collected and translated by Ofer Fridman. They offer a very different perspective on familiar issues and themes. Western strategists may not turn to them for guidance, but they provide insights into a tradition that may be reviving in modern Russia, and they certainly make you think.

Lawrence Freedman
February 2021

EDITOR'S INTRODUCTION

This book is not intended exclusively for experts on Russian affairs. Though they might find it useful, they can easily access the original texts, which have been recently digitalised by the Russian State Library. Instead, it aims to attract the attention of those who are interested in understanding Russia's strategic behaviour, but do not have the required language skills to access the texts presented in this book.

This project, like many other similar projects, started with a conversation about the importance of cultural-historical context for the understanding of states' strategic behaviour. While this conversation touched upon many different arguments, it finally gravitated towards one 'simple' question: can we understand contemporary Western strategy without reading Clausewitz? The answer of my senior colleague was an unambiguous no. 'Not that we have to agree with him,' she continued, 'but Clausewitz represents such an important pivot of the Western understanding of strategy that, without reading his work, you will get lost in the forest of references and allusions to it.'

As someone who researches modern-day Russian military thinking, I could not stop myself from wondering: is this true also for Russia? While the immediate answer to this question was yes—Clausewitz has been as influential in Russia as he is in the West (perhaps even more so)—my journey through the deep waters of Russian strategic thinking brought me to quite unexpected shores. While modern Western literature generally confines its analysis of Russian strategic thinking to the Soviet period only, prominent works of current Russian strategists argue that the Russian understanding of war has much deeper and wider historical-cultural roots.

STRATEGIYA

With the collapse of the Soviet Union and the decline of the Russian military physically (due to the division of the Red Army between fifteen new states) and intellectually (due to the loss of Soviet ideology), Russian military thinkers began looking for a new philosophy of war. As a pair of Russian officers stated in 1994: 'By rejecting the exclusive role of Marxism as the sole true teaching that explains the nature and character of war [...] we face the need to clarify the [military] scientific basis of [our] worldview, our views of war as a special societal condition.'[1]

During the time when Russian politics began looking towards the West, trying to absorb and adapt Western practices, Russian strategy did not follow the same direction. Instead, it went on a long journey of rediscovery of the Russian traditional school of strategic thinking. For example, *Military Strategy*, the 2003 seminal work[2] of Sergey Mikhalev, lists different definitions of strategy that 'illustrate the development of the understanding of this category of military art'.[3] This chronological list includes definitions by the following thinkers: Dietrich Heinrich von Bülow; Archduke Charles, Duke of Teschen; Carl von Clausewitz; Peter Alexandrovich Yazykov; Antoine-Henri Jomini; Genrikh Antonovich Leer; Helmuth von Moltke the Elder; Evgeny Ivanovich Martynov; Sigismund von Schlichting; Nikolai Petrovich Mikhnevich; Aleksei Yevgenievich Gutor; Alexander Andreyevich Svechin; Vasily Danilovich Sokolovsky; Nikolai Vasilyevich Ogarkov; and Andrian Aleksandrovich Danilevich.[4]

While this list, by itself, already says a lot to any keen student of strategy, there are three main observations that must be highlighted for a reader who is less familiar with the topic. Firstly, there is the visible absence of the ideologists of Soviet strategy—Vladimir Lenin and Mikhail Frunze. Even Marshal Georgy Zhukov, the greatest warrior of the Soviet Union, is not on the list. Secondly, the most recent non-Russian military thinker who should, according to Mikhalev, inspire future generations of Russian strategists is Sigismund von Schlichting, a Prussian general and military theorist, whose career peaked in the last two decades of the 19th century. In other words, the whole Western strategic school of the 20th century is not on the list either. This does not imply that Russians are unfamiliar with the Western works on strategy. On the contrary, their books are full of references to these

EDITOR'S INTRODUCTION

works. It simply suggests that they consider them for what they are—a Western interpretation of strategy that does not necessarily represent the Russian school of strategic thought.

The third observation, and probably the most important for this book, is that out of nine Russian military thinkers on this list, the selected works of only two have been translated into English: Svechin's *Strategy*[5] and Sokolovsky's *Military Strategy*.[6]

These observations ultimately led me towards two important questions. The first question was: is the Russian interpretation of strategy similar to that of the West? Interestingly enough, there is no simple answer to that question. On the one hand, Russia's strategic school, like its Western counterpart, is rooted in the interpenetration of the Prussian (Bülow and Clausewitz), French (Jomini) and British (Henry Lloyd) strategic schools of the late 18th and first half of the 19th century. On the other hand, despite the same source, the Russian river of strategic thinking, from its very beginning, took its own route, following its way through the vast wilderness of Russian territory (and not the highly populated and industrialised Europe), being coloured by the values of the Russian Orthodox Church (and not by Catholicism), passing the rapids of Communist ideology (and not Western democracy and capitalism), and finally finding itself muddied after plunging over the waterfalls of Russian state collapse (twice in one century). The Volga is still a river, but, with its own characteristics, it is different from the Thames, Rhine, Seine or Mississippi.

The same is true of present-day Russian strategic thought. For example, the Russians have never seriously discussed the idea of grand strategy (*bol'shaya strategya*), simply describing it as a Western concept that they should be aware of but not necessarily develop further.[7] Instead, the closest contemporary alternative to the Western concept of grand strategy is national strategy (*natzional'naya strategya*).[8] Another example is that the Russian strategic lexicon does not have the word 'deterrence'. Referring to nuclear or conventional deterrence, Russians use the word *sderzhivaniye*, which literally translates as containment. While one might say that these are just semantics, as each side knows what the other implies, this is not entirely correct. These differences provide the context that makes Russian strategy (its conceptualisation and implementation) truly Russian.

STRATEGIYA

According to Lawrence Freedman, context helps to explain the power of theory, as well as its limitations.[9] In other words, without understanding the context, any assessments, judgements or critiques of a theory (as well as the outcomes of its implementation) venture on the slippery slope towards misjudgement and misinterpretation. For example, in the last decade, some Western experts have portrayed the actions of President Putin (and Russia in general) as astrategic and tactical.[10] While the Kremlin's recent actions might not follow what Freedman defines as 'strategic behaviour',[11] this does not necessarily mean that these actions contradict the traditional Russian understanding of strategy.

In Russian strategic eyes, free of Western attempts to deconstruct strategy into mechanical elements and procedures (ends, means and ways), strategy has always remained a work of art. From Genrikh Leer ('a pioneer of Russian strategic thought'),[12] through Alexander Svechin (whose 'timelessly vital conceptual school finds nowadays more and more supporters')[13] and Nikolai Ogarkov ('who creatively comprehended new problems of military art'),[14] to the contemporary dictionary of the Russian Ministry of Defence, strategy has always been understood as 'an integral part of the art of war'.[15]

Therefore, Russian strategy, like any work of art, cannot be measured. It can only be appreciated. Putin is now entering his third decade of power. Along the way, he brought his country from the economic and political chaos of the 1990s back onto the global international stage. He successfully (at least from the perspective of the Kremlin's understanding of Russia's national interests) waged few wars, forced the West back to the shelters of the Cold War (if not physically, then intellectually), befriended China (the second-largest economy in the world), and became a welcome guest not only in the capitals of countries at odds with the West, but also at the wedding of the Austrian foreign minister. This is not just a work of art, it is a masterpiece of strategy. It might not be to Western taste, but you do not have to like Dostoyevsky to appreciate his work. As Anton Kersnovski ('one of the brightest figures of the military culture of the Russian émigré')[16] wrote: 'art [...] is a national affair. Regardless of the field of art (whether it is literature, painting or military affairs), nationality is its most characteristic aspect, its so-called "flavour", its quintessence'.[17]

EDITOR'S INTRODUCTION

This leads me to the second question that drove me to undertake this project: who are these Russian Clausewitzes, Jominis and Liddell Harts who have been inspiring current Russian strategic thought? Having read through many contemporary Russian works on strategy,[18] and given special attention to the references provided by their authors, I can argue that the contemporary Russian understanding of strategy is rooted in three main groups of literature: the classic works of the 19th-century Russian strategists; the analytical literature produced by Russian military thinkers in exile; and various prominent works of Soviet strategists. While some works of Soviet military thinkers have already been translated into English,[19] two other schools of Russian strategy have remained inaccessible to a Western reader interested in understanding present-day Russian strategic behaviour. Therefore, tasked with the aim of filling this lacuna, this book offers a view of the foundations of the Russian art of strategy.

How to Read This Book

When I started this project, one of its earlier critics told me, 'Why are you doing this? It is not as if the Russians use these books as manuals!' Well, these are not manuals indeed, and the reader should be warned from approaching them as such. Considering them as manuals would be similar to thinking of Clausewitz's *On War* as a manual for present-day Western warfare. If strategy is an art, then the purpose of the masterpieces of the past is to inspire the readers, and not tell them what to do. And these works do exactly that.

For example, the founding father of Russian strategic school, Genrikh Leer, who was rediscovered in the 1990s, stated that the main goal of strategy is 'to grasp the question of waging war at a given moment in all its aspects and solve it according to the prevailing situation, i.e., to define a reasonable goal and direct all forces and means towards its achievement in the shortest time and with the least sacrifices'.[20] One of his best students, Evgeny Martynov (according to Kokoshin, 'a distinguished Russian Imperial and Soviet military theoretician'),[21] applied this understanding well to Soviet strategy, claiming that politics should enable strategy to 'achieve the desired result in the shortest time and with the minimum of costs'.[22] If this is the inspiration behind modern-day Russian strategy, accusing it of astrategic behaviour is similar to accusing a painter of being an Impressionist and not a Realist. We may favour Gustave Courbet over Claude Monet, but we must acknowledge that both were very talented artists who have inspired generations of painters.

Following this logic, this book presents selected works from two schools of Russian strategic thinking that have re-emerged after the fall of the Soviet Union: Imperial and Russian-in-exile. Since the early 1990s, these works have been widely republished, analysed and commented on by the Russian strategic community. Moreover, as stated above, the works of current Russian strategists do not have fewer references to these works than to the writings of Soviet strategists (perhaps even more). In other words, when following the collapse of the Soviet Union 'all previously accepted military-political and strategic guidelines collapsed in front of our eyes',[23] these works provided the required inspiration that has allowed Russia's strategy to recover.

The Russian Imperial school is represented in this book by the works of General of the Infantry Genrikh Leer (1829–1904), General of the Infantry Nikolai Mikhnevich (1849–1927) and Lieutenant General Evgeny Martynov (1864–1937). The curation of the Russian Imperial strategic art is not an easy task. The Russian Imperial school of strategic thinking included many talented strategists—General of Artillery Baron Nikolai Medem[24] ('the first Russian professor of strategy'),[25] Field Marshal Dmitry Milyutin[26] (who served as Russia's Minister of War from 1861 to 1881 and was responsible for a series of fundamental reforms that reshaped Russia's military organisation), General of the Infantry Mikhail Dragomirov[27] (who was the first to translate Clausewitz's *On War* into Russian (from French) in 1888, and whose original works 'are rightfully considered the classics of the Russian military school'),[28] and many others.

The works of Leer, Mikhnevich and Martynov were chosen for a number of reasons. Firstly, these works focus on strategy, and not on tactics (like Dragomirov's most influential treatise, *Textbook on Tactics*, which 'for over twenty years served as the primary training manual for officers in the art of tactics')[29] or military history (like most of the works of Milyutin). While it is difficult to disagree with Christopher Duffy, who argued that 'every [present-day] historian of the old Russian army

EDITOR'S INTRODUCTION

has lived parasitically off the mighty product of the Russian scholars of the second half of the nineteenth century',[30] the focus of this book is on Russian strategic thinking, and not military historiography.

The second and most important reason for the choice of Leer, Mikhnevich and Martynov is their influence on current Russian strategic thinking. Unlike the works of their counterparts, the works on strategy by Leer, Mikhnevich and Martynov have been widely republished in present-day Russia, restoring the discourse on the nature of strategy.[31] Since the purpose of this book is not to objectively present the state of strategic thought in Imperial Russia, but emphasise its influence on Russia's present-day strategic thinking, it seems obvious to focus on the works that current Russian strategists find useful, and omit the rest.

Genrikh Leer was chosen because he is widely considered to be a founding father of Russian strategic thinking, whose works 'have similar significance to the works of Henry Lloyd, Antoine-Henri Jomini and Carl von Clausewitz'.[32] Leer made a valuable contribution to the development of the art of war and strategy in Russia, dividing it into theoretical and practical parts. The most influential outcome of his research was *The Experience of Historical-Critical Research into the Laws of Military Art (Positive Strategy)*, in which he not only set the course for the Russian ship of independent strategic thinking, but also equipped it with the required methodological sails.

The works of Nikolai Mikhnevich and Evgeny Martynov were selected for two reasons. Firstly, together with Leer, they represent three successive generations of the Russian Imperial school of strategic thinking, from its maturity in the second part of the 19th century until well into the early period of the Soviet Union. Mikhnevich and Martynov were good students of Leer, adopting his methods and developing them further, reinforcing the unique Russian way of strategy. For example, reading Martynov's *The Responsibilities of Politics in Its Relations with Strategy*, it is difficult not to notice his use of Leer's method of critical-historical analysis, or his attempt to answer Leer's challenge that 'an analysis of the influence of the political conditions on waging wars [...] should constitute the field of a separate, though still non-existing, military science—*military politics*'.[33] In addition, Mikhnevich's attempt to distinguish between the plan of war and the

7

plan of campaign in his *Foundations of Strategy* leaves no doubt about the profound influence of Leer.

The second reason for the choice of Martynov and Mikhnevich was that both continued to influence Soviet thinking after the 1917 Bolshevik Revolution. The Red Army, as well as its strategic thinking, contrary to the assumption of some Western experts, was not created from scratch. Many Russian Imperial generals continued to serve in the Red Army until Stalin's Great Purges in the late 1930s, including the influential Alexander Svechin, who was a major general in the Imperial Army and a great student of Mikhnevich.[34] When some Western commentators see in present-day Russian behaviour the Soviet 'active measures',[35] I tend to direct them towards Matynov's 1899 *The Responsibilities of Politics in Its Relations with Strategy*:

> Almost every country carries within it the germ of internal political or social disease. Therefore, in addition to the external threat, it might have internal ones that subvert its power. Good politics should find them by scrupulously studying the enemy's governmental structure, social life and the ruling classes of its people. [...] In exploiting these weaknesses of the enemy's state, good politics will always find collaborators who are dissatisfied with the existing order.[36]

In other words, the strategic idea of subversion has much deeper roots in Russian strategic thinking, which interweaves the Imperial, Soviet, Russian-in-exile[37] and current schools of strategic thought. (This idea does not seem to be uniquely Soviet or Russian, judging by the American Revolution, which was supported by the French; the Arab Revolt, which was orchestrated by the British; the Bolshevik Revolution, which was instigated by the Germans; and the 1953 Iranian coup d'état, which was staged by the Americans.)

Another important characteristic of strategy that has much deeper roots in Russian strategic thinking than is commonly assumed is the idea of combination. The idea of so-called 'Russian hybrid warfare', which combines 'the use of military and non-military tools in an integrated campaign', has since 2014 preoccupied Western conceptual discourse,[38] accusing the current Kremlin of blurring the line between war and peace.[39] The common problem of contemporary Western analysis of Russian strategic thinking is that it approaches it as either a transformation of or a transition from Soviet strategy, ignoring the

EDITOR'S INTRODUCTION

Russian Imperial school of strategy or the influence of Russian military thinkers in exile.

For example, one of the most important characteristics of the Russian Imperial school of thinking is its rejection of the universality of strategy and its emphasis on the importance of the prevailing situation in strategy-making. As Leer put it: 'There are no laws (*rules*) that suit every possible occasion, because the number of possible occasions is infinite. [...] Therefore, each law should be seen as a formula, in which situation (*time* and *occasion*) is a variable that should be inserted to get a specific solution to every known occasion.'[40] Thirty years after Leer, Martynov persisted in the same direction, arguing that '*the methods of strategic art* usually change with the appearance of a new situation'.[41] Therefore, Russian Imperial strategists repeatedly emphasised the importance of *glazomer* (*coup d'œil*—literally, an ability to measure distances by eye), which, as Mikhnevich stated, 'was regarded by Suvorov as the first quality required of a military commander'.[42] Leer's definition of *glazomer* as 'a continuous accurate assessment of the situation, time, conditions and space required to achieve one or other combination'[43] leads to another fundamental characteristic of the Russian Imperial understanding of strategy—the importance of combination in strategy-making.

As Medem put it in 1836: 'all great commanders were truly great because they based their actions not on pre-existing rules, but on a skilful combination of all means and circumstances'.[44] In other words, Russian strategists tended to believe that strategy should not follow any specific rules. Instead, they insisted that it represented an artful combination[45] of strategic elements (moral, geographical, tactical, administrative, political, and the element of chance)[46] that answered the specific requirements of each given situation. If this is the inspiration behind present-day Russia's strategy-making, then we should not be surprised that the Kremlin does not subscribe to the West's idealised bins of strategy (ends, means, ways) or pristine distinctions between military and non-military tools.

This leads us to the school of Russian military thinking in exile, which is represented by the works of Anton Antonovich Kersnovski (1907–44), Lieutenant General Nikolai Nikolaevich Golovin (1875–1944) and Colonel Evgeny Eduardovich Messner (1891–1974). As with the selection process for the Russian Imperial texts, it was not easy to choose works to represent the Russian strategic school in exile. Therefore, this process was guided by similar intentions. Firstly, the focus was given to works on strategy and not military history. For example, many works produced by Russian military thinkers in exile focused on the First World War and the Bolshevik Revolution with two main goals in mind: providing historical accounts based on personal experience and offering analytical insights into the causes behind these events. One of the most influential examples of this type of historical-bibliographical work is *The Russian Turmoil* by Lieutenant General Anton Denikin.[47] While this work is indispensable for an understanding of the political, social and military events between 1917 and 1920 that led to the fall of the Russian Empire and the defeat of the White Movement in the Civil War, it offers little conceptual-theoretical contribution to the understanding of strategy. The second reason for choosing the works of Kernovski, Golovin and Messner is that, like Leer, Mikhnevich and Martynov, they have been widely republished in present-day Russia, shaping (as many of the thinkers in exile hoped and envisioned) the discourse on the nature of strategy in post-Soviet Russia.[48]

On the one hand, until the collapse of the Soviet Union, Communist ideology restricted access to the Russian émigré military and political thinkers, who were determined anti-Communists. On the other hand, the disappearance of these restrictions, combined with the thirst for a new perspective on the phenomenon of war, served as fruitful soil for the ideas formulated by the Russian strategic school in exile, which took a different path from that of the Soviets after the end of the Civil War. Colonel Alexander Savinkin, the founder and main editor of *Rossiyskiy Voyennyy Sbornik* [Russian military collection]—a series supported by the Russian Military University and the Russian Ministry of Defence—has edited numerous books that republish and analyse the works of Russian émigré military and political thinkers. In his words:

> During their lifetime, the White émigrés did not and could not solve their political tasks, but they left a great spiritual legacy, values and guidelines for the development of a patriotic education for the future of the armed forces of liberated Russia. The spiritual revival of the

EDITOR'S INTRODUCTION

military today is [thus] impossible without studying and understanding the patriotic and military thinking of the Russian émigrés, the ideas of the White [Movement] and [its] traditions.[49]

Reading the works of these thinkers is not an easy task, for two main reasons. The first is that, for most of them, the Civil War was not lost on the battlefield, but in the hearts and minds of the Russian people, a battle lost due to poor political-military leadership. The second reason was best formulated by Igor Domnin, Savinkin's frequent co-editor and co-author, who argued that the military thinkers in exile 'created their valuable military-societal works independently, without any official support' and, as a result, these works are characterised by 'an unprecedented freedom of military thought (more accurately, freedom of expression of thought), its uncensored nature, suppleness, complete theoretical and ideological pluralism', which 'inscribed their works with maximum criticism and revelation'.[50] In other words, many of them published by themselves and for themselves, and, as one critic of Kersnovski's *Philosophy of War* put it: 'his book is neither a handbook nor a work of academic research. It is neither dry enough to be the former, nor is it correct enough to be the latter. [...] In his style of presenting events and explaining them, Kersnovski resembles a virtuoso pianist who does not look at the keys as they run under his fingers.'[51]

Therefore, the first work presented here is Kersnovski's *Philosophy of War*, published in 1939 as a collection of previously published articles. I would like to warn the reader against judging this work on its accuracy or academic rigour. Instead, I would suggest we appreciate its inspirational power on the hearts and minds of Russian strategists in the 1990s who, like Kersnovski, found themselves amid the ruins of Russian power and blamed the political leadership for leading Russia's military and state to the abyss. Hence, I would suggest that the reader try to appreciate the reassurance and inspiration which the bemused Russian strategists of the 1990s (watching their president, Boris Yeltsin, dancing on TV) found in Kernovski's words:

> The pathological era of 1914–18 was followed by the period of collective softening of the brain (the consequence of the serious concussion caused to the world by this war). This period is called the era of 'democracy'. The foreign policy consequence of this epoch was the replacement of the old cabinet school of diplomacy by a new one—the

school of carnival. The professionals of state affairs were replaced by amateurs. Knowledgeable people were replaced by ignorant speakers used to addressing rallies, who held the title of 'people's representatives', but did not always have even a primary school certificate.[52]

The second work of the Russian émigrés presented in this book is Golovin's *The Science of War*, published in 1938. Golovin represents an entirely different side of Russian military thinking in exile. While Kersnovski is typical of those who focused on the question of 'how and why we lost our Fatherland', Golovin represents those who focused on the meaning of what had happened during the First World War (including the Bolshevik Revolution) for the art and science of war. In other words, being 'a distinguished Russian military scientist' (according to present-day strategist Andrei Kokoshin),[53] Golovin tried to develop this science further.

On the one hand, Golovin's work presented here is not an easy read. Heavily based on John Stuart Mill's *A System of Logic* and other philosophers, it tries to justify and promote Golovin's idea of a 'sociology of war' which 'not only would offer a rationale for the theory of military art, but it would also define the boundaries within which the application of theoretical generalisations is valid, i.e., scientific.'[54] On the other hand, this work represents the most interesting and intriguing link in the development of strategic thinking both in Russia and in general.

Firstly, in 1900 Golovin finished his studies at the Nicholas General Staff Academy, where he was influenced by Genrikh Leer (who was the director of the Academy until 1898) and Nikolai Mikhnevich (who was a professor from 1892 and the director from 1904 to 1907). This influence can easily be traced in Golovin's writing not only in his references, but also in his ideas. For example, whereas Leer stated that 'while military psychology is a science of the distant future, it is still important to pursue this field',[55] Golovin approached this challenge seriously, arguing that the conditions had become ripe for the creation of military psychology as 'an auxiliary science at the service of the sociology of war'.[56] Another example is the very idea of the 'sociology of war', which, as Golovin acknowledges himself, was initially formulated by Mikhnevich.[57] Therefore, Golovin's work represents a good example of the continuation of thinking that originated in the mid-19th century.

EDITOR'S INTRODUCTION

Secondly, in analysing present-day Russian understanding of the role and place of war, it is difficult not to notice the influence of Golovin's ideas. He diverged from the thinking of his counterparts, who stayed in Soviet Russia, and took his own route. However, with the collapse of the Soviet Union, his idea that an analysis of war 'should be the examination of war as a phenomenon of social life, and not only an analysis of ways to wage wars'[58] found its way back to Russia (Kokoshin's book *The Political Science and Sociology of Military Strategy* provides a good example of this).

This leads to the third and most intriguing aspect of Golovin's work. In pursuit of his idea of establishing the sociology of war as a branch of the social sciences, he knocked on every door, visited every capital and presented his ideas at every forum that was ready to listen. However, in inter-war Europe, which was struggling to recover (materially and spiritually) from the Great War, nobody was ready to take a Russian general seriously and accept his argument that this terrible war was part (and product) of social life. It took another world war for the scientific community to accept it and start studying the phenomenon of war in its broader social-cultural context. However, it is difficult for me, as a member of the Department of War Studies at King's College London (the first academic institution to address this field), not to appreciate the prescience of Golovin's words:

> I completely understand the difficulties attached to the establishment of [a] special scientific institute for the sociological study of war. However, I believe that, sooner or later, humanity will get there. Meanwhile, I will allow myself to express a humbler request—the establishment of a department of the 'sociology of war'. Only the establishment of this department, at least at one university that has a faculty of social sciences, can be considered as the first true attempt to scientifically study war as a phenomenon of human social life.[59]

The final chapter of this book presents Evgeny Messner's *The Face of Contemporary War*, published in 1959 in Buenos Aires. Messner represents a younger generation of military thinkers in exile, who fought against the Bolsheviks twice (in the Civil War and in the Second World War) and lived through the first half of the Cold War. Together with other Russian émigré military and political thinkers, his works were rediscovered and popularised after the collapse of the Soviet Union, becoming one of the conceptual foundations of modern-day Russian understanding of political subversion. Since Messner's work is, probably, the most influential among all those presented in this book, I would somewhat egoistically refer the reader to my previous book which discusses his influence in detail.[60]

While the levels of conceptual rigour and historical accuracy vary across the works of Russian military thinkers in exile, their contemporary counterparts find in this literature two very important, interconnected lessons. The first is about the phenomenon of war. While the psychological dimension of war had already been discussed during the most productive period of Russian military thinking in exile (the 1920s and 1930s), 'during its later period (1946–60), Russian military thought in exile unequivocally determined that modern warfare reached beyond the usual three environments—of land, water and air—into a fourth—"the soul of the soldier, the soul of belligerent nations" (from three types of confrontation—economic, diplomatic and physical—to a fourth—psychological)'.[61]

Returning to the argument presented by some Western experts about Soviet 'active measures' as a conceptual lens through which to examine contemporary Russia's strategic behaviour, it seems that this lens offers a very limited understanding. While there are undeniable similarities between Soviet 'active measures' and present-day ways and means used by the Kremlin, an analysis of the conceptual debate on political subversion in Russia shows that the works of the military thinkers in exile (who believed that they lost their fatherland to Bolshevik subversion) have significantly wider influence than those of their Soviet counterparts (who failed, according to the post-Cold War Russian strategists, to protect their fatherland against the subversion of the West).[62] After all, already in 1959, Messner argued that contemporary war was about 'degrading the spirit of the enemy and saving your own spirit from degradation'.[63]

The second lesson, which contemporary Russian strategists find useful in the works of the military thinkers in exile, concerns the rebuilding of Russia after the collapse of the Soviet Union, and of its armed forces after the unsuccessful campaigns in Afghanistan in the 1980s and Chechnya in the first half of the 1990s. The situation in which Russian strategists found themselves in the 1990s was quite desperate. As James

EDITOR'S INTRODUCTION

Pearce puts it, 'this "decade without patriotism" was symbolic of deep pessimism and lack of hope'.[64] Attempting to address this lack of patriotism and demoralisation in the armed forces was very important, especially in the light of an understanding of the increasing role of the psychological dimension of war. The writings of Russian military émigrés, who 'devotedly and selflessly served their Fatherland with both sword and pen',[65] proved especially useful for this purpose, as Domnin describes it:

> In their work, the military émigrés paid special attention to the element of the spirit: the education, moral-psychological state, and ethical and political health of the army. The words of these exiles were nothing more than a paternal lesson from the last generation of the Russian Imperial Army about its spiritual experience and its soul. It is a bridge of continuity from our 'yesterday' to our 'tomorrow'.[66]

To conclude this short introduction, I would like to repeat my warning about the purpose and nature of this book. According to the editors of a similar compilation published in Russian, 'in studying the legacy of Russian officers, we must focus on its substance, using it to strengthen our spirit and solve modern problems'.[67] Similarly, I advise the reader to follow the substance and ideas presented in the translated works, and not their letter, while trying to appreciate their influence and not cavil about their weaknesses.

A Note about Translation, Punctuation and Transliteration

Russian, like any other language, is deeply embedded in its own unique historical-cultural context, and, like any translator, I have had to navigate between following the words verbatim or trying to translate their meaning. However, sometimes the intricate particularities of a language simply prevent the translator from conveying the true meaning of a given text. *Russkiy* and *rossiyskiy*, for instance, are both translated into English as 'Russian', but there is a vast difference between *russkaya zemlya* (Russian land) and *rosiiyskaya federatziya* (Russian Federation) because *russkiy* describes something that belongs to the Russian ethnos, whereas *rossiyskiy* describes something belonging to the Russian state.

Attempting to convey the meaning of a text, rather than providing a direct translation, can prove dangerous, as the translator will inevitably run the risk of putting his own words into the original author's mouth. But a direct, word-for-word translation is no less dangerous, as it risks losing the meaning altogether. Ultimately, despite the risks involved, I tried to navigate carefully between these two approaches, and I take full responsibility for any mistakes and misinterpretations. My only hope is that my awareness of the pitfalls of both approaches has kept my mistakes to a minimum.

For whatever reason, Russian writers are very parsimonious when it comes to references. Moreover, during the 19th century there was a practice of paraphrasing quoted material (making it almost impossible to identify the sources), while still attributing it to the original author. For example, Leer and Mikhnevich frequently provide so-called 'popular sayings' (*krylatye vyrazheniyia*) of Clausewitz, Napoleon and other military thinkers, without necessarily providing a reference to where and when they were said or written. In places where the source could be identified (either because of the substantial length of the quote, or the popularity of the quoted statement), I relied on the original English text, referring the reader to the source. On other occasions, unable to find the original quote, I simply translated it from the Russian.

Therefore, all footnotes to the translated texts are mine, and they seek to help the reader identify Russian historical figures, events from Russian history, and the various sources used by the authors. All original footnotes with explanations and references provided by the authors are placed within the text in braces or curly brackets {—}. All emphases within the texts, either in *italics* or **bold**, follow the original texts. Owing to the limitation of space, the works are not presented in their complete form. Instead, I had to take the very difficult decision to select the parts that seemed to be particularly valuable, leaving the rest out. The omitted parts are indicated by [...].

Many contemporary researchers have warned that the West's tendency to project its own concepts onto Russian thinking is often misleading, offering a very partial insight into Russian conceptions of strategy and its implementation.[68] To signal my assent to this important argument, I have decided to transliterate the Cyrillic script according to the Russian standard GOST 7.79–2000, and not according to the Library of Congress rules, which are widely accepted in the West. Following George Orwell's assessment, 'if the thought corrupts lan-

EDITOR'S INTRODUCTION

guage, language can also corrupt thought',[69] I believe that we have a better chance of truly understanding Russian *стратегия* once we start seeing it as *strategiya* (according to GOST 7.79–2000) and not as *strategiia* (according to the Library of Congress). After all, this is how Russians see it.

Finally, while I do not necessarily subscribe to the arguments and views presented by the authors, the responsibility for all mistakes, misspellings and other linguistic faults is mine and mine alone, and I offer the reader my humble apologies for this.

NOTES

1. V. Solov'ev and A. Dremkov, 'Yeshche raz o predmete i strukture voyennoy nauki' [Once again on the subject and structure of military science], *Voyennaya Mysl'*, no. 9, 1994, pp. 34–5.
2. Andrey Kokoshin, one of the renowned Russian contemporary strategists, describes this book as 'authoritative', 'fundamental' and 'capital'. See Andrey Kokoshin, *Politologiya i sotsiologia voennoi strategii* [The political science and sociology of military strategy], (Moscow: URSS, 2016), pp. 20, 37, 55.
3. Sergey Mikhalev, *Voennaya strategia: Podgotovka i vedenie voyn novogo and noveyshego vremeni* [Military strategy: Preparing for and waging wars in modern and contemporary history], (Moscow: Kuchkovo Pole, 2003), p. 24.
4. Ibid., pp. 22–23.
5. Alexander Svechin, *Strategy*, edited by Kent D. Lee (Minneapolis: East View, 1993).
6. Vasily Sokolovsky (ed.), *Military Strategy: Soviet Doctrine and Concepts* (London: Pall Mall Press, 1963).
7. See Alexander Vladimirov, *Osnovy obschey teorii voyny—Chast' II: Teoriya natsional'noy strategii* [The foundations of the general theory of war—Part II: The theory of national strategy], (Moscow: Moskovsii Finansovo-Promyshlennyy Universitet 'Sinergiya', 2013); Valeri Konyshev and Alexander Sergunin, *Soveremennaya voennaya strategiya* [Contemporary military strategy], (Moscow: Aspect Press, 2014); Kokoshin, *Politologiya i sotsiologia*.
8. Ibid.
9. Lawrence Freedman, 'Author's Response by Lawrence Freedman, King's College London', *H-Diplo: ISSF Roundtable*, vol. 10, no. 14, 2018, pp. 18–21.

10. For example Michael A. McFaul, 'The Myth of Putin's Strategic Genius', *New York Times*, 23 October 2015, https://www.nytimes.com/2015/10/23/opinion/the-myth-of-putins-strategic-genius.html (accessed 20 October 2020); Joshua Rovner, 'Dealing with Putin's Strategic Incompetence', *War on the Rocks*, 12 August 2015, https://warontherocks.com/2015/08/dealing-with-putins-strategic-incompetence/ (accessed 20 October 2020).
11. See Lawrence Freedman, *Ukraine and the Art of Strategy* (Oxford: Oxford University Press, 2019).
12. Peter Von Wahlde, 'A Pioneer of Russian Strategic Thought: G.A. Leer, 1829–1904', in Roger Reese (ed.), *The Russian Imperial Army 1796–1917* (London: Routledge, 2017). I give this reference to a source in English, only for the purpose of making it accessible to the reader.
13. Andrei Kokoshin, 'Predislovie' [Foreword], in Aleksander Savinkin (ed.), *Postizheniye voyennogo iskusstva: Ideynoye naslediye A. Svechina* [Comprehending military art: The ideological heritage of Aleksander Svechin], (Moscow: Russkii Put', 2000), p. 17.
14. Makhmut Gareev, '"Sem'" let iz zhizni marshala' [Seven years from the life of the Marshal], *Voenno-Promyshlenyi Kur'er*, no. 48 (4665), 5 December 2012.
15. Ministry of Defence of the Russian Federation, 'Strategiya' [Strategy], in *Voennyy entsiklopedicheskiy slovar'* [Military Encyclopedic Dictionary], http://encyclopedia.mil.ru/encyclopedia/dictionary/list.htm (accessed 20 July 2020).
16. Igor Domnin, 'Gorenie dukkha: Slovo ob Antone Kersnovskom' [Burning spirit: A word on Anton Kersnovski], in A.B. Grigor'ev (ed.), *Filosofiya voyny* [Philosophy of war], (Moscow: Izdatel'stvo 'Ankil-Voin', Rossiiskiy Voennyy Sbornik, 1995), p. 205.
17. Anton Kersnovski, *Filosofiya voyny* [Philosophy of war], (Belgrade: Izdanie 'Tzarskogo Vestnika', 1939), p. 22.
18. Since listing all relevant sources is impractical, I refer the reader to the most fundamental works that served as a starting point of my research. To the works already mentioned of Andrey Kokoshin, Sergey Mikhalev and Alexander Vladimirov, I would like to add Vasily Mikryunov, *Voyna: Nauka i iskusstvo* [War: Science and art], 4 vols. (Moscow: Rusayns, 2016), S. Yu. Tyushkevich, *Otechestvennaya voennaya nauka: Stranitsy istorii, problem, tendentsii* [Fatherland's military science: Pages of history, problems, tendencies], (Krasnodar: Krasnodarskii Yuredicheskii Institut MVD Rossii, 2001).
19. For example: Vladimir Triandafillov, *The Nature of the Operations of Modern Armies*, translated by William Burhans, edited by Jacob Kipp

EDITOR'S INTRODUCTION

(London: Routledge, 1994); Georgii Isserson, *The Evolution of Operational Art*, translated and edited by Bruce Menning (Fort Leavenworth: Combat Studies Institute Press, 2013); Svechin, *Strategy*; Sokolovsky, *Military Strategy*. Andrey Kokoshin discusses knowledgeably the Soviet strategic school in his *Soviet Strategic Thought, 1917–91* (Cambridge: MIT Press, 1998).

20. Genrikh Leer, *Opyt kritiko-istoricheskogo issledovaniya zakonov vedeniya voyny (polozhitel'naya strategiya)*, [The experience of historical-critical research into the laws of military art (positive strategy)], (St Petersburg: Pechatnya V. Golovina, 1869), pp. 23–24.
21. Kokoshin, *Politologiya i sotsiologiya*, p. 71.
22. Evgeny Martynov, *Obyazannosti politiki po otnosheniyu k strategii* [The responsibilities of politics in its relations with strategy], (St Petersburg: Tipografiya Glavnogo Upravleniya Udelov, 1899), p. 1.
23. Adrian Danilevich, 'Sovremennaya voennaya strategiya Rossii i perspektivy yeye razvitiya' [Contemporary military strategy of Russia and the prospects for its development], in Vladimir Zolotaryov (ed.), *Istoriya voennoy strategii Rossii* [The history of Russian military strategy], (Moscow: Kuchkovo Pole, 2000), p. 497.
24. Nikolai Medem, *Obozrenie izvestniyshikh pravil i system strategiy* [An overview of the most famous rules and systems of strategy], (St Petersburg: Tipogragiya II Otdeleniya Sobstvennoy E.I.V. Kantzelyarii, 1836).
25. Alexander Svechin (ed.), *Strategiya v trudah voennyikh klassikov* [Strategy in the works of military classics], vol. 1 (Moscow: Vysshiy Voennyi Redaktsionnyi Sovet, 1924), p. 85; also Marina Yeliseeva, 'Pervyi otechestvennyi professor strategii' [The fatherland's first professor of strategy'], *Krasnaya Zvezda*, 3 April 2019, http://redstar.ru/pervyj-otechestvennyj-professor-strategii/ (accessed 20 October 2020).
26. For example: Dmitry Milyutin, *Kriticheskoye issledovaniye znacheniya voyennoy geografii i voyennoy statistiki* [A critical study of the significance of military geography and military statistics], (St Petersburg: Voennaya Tipografiya, 1846); *Opisaniye voyennykh deystviy 1839 goda v Severnom Dagestane* [The description of military operations in 1839 in Northern Dagestan], (St Petersburg: Tipogragiya Voenno-Uchebnykh Zavedenii, 1850); *Istoriya voyny Rossii s Frantsiyey v tsarstvovaniye Imperatora Pavla I v 1799 godu* [The history of the war between Russia and France during the reign of Emperor Paul I in 1799], 5 vols. (St Petersburg: Tipogragiya Voenno-Uchebnykh Zavedenii, 1852–53).
27. For example: Mikhail Dragomirov, *Razbor romana 'Voyna i Mir'* [An analysis of the novel 'War and Peace'], (Kiev: Izdanie Knigoprodavtsa N. Ya. Ogloblina, 1895); *Opyt rukovodstva dlya podgotovki chastey k boyu* [Leadership experience in preparations of units for battle], vol. 1 (Kiev:

Tipografiya Okrughnogo Shtaba, 1870), vol. 2 (Kiev: Tipografiya Okrughnogo Shtaba, 1871), vol. 3 (St Petersburg: Tipografiya Trenke i Fyusno, 1886); *Uchebnik taktiki* [Textbook of tactics], (St Petersburg: Tipografiya V.S. Balasheva, 1879).

28. Aleksey Mazur, 'K 185-letiyu so dnya rozhdeniya M.I. Dragomirova' [On the 185th anniversary of the birth of M.I. Dragomirov], Ministry of Defence of the Russian Federation, http://энциклопедия.минобороны.рф/encyclopedia/history/more.htm?id=12070899@cmsArticle (accessed 20 July 2020).
29. Ibid.
30. Christopher Duffy, *Russia's Military Way to the West: Origins and Nature of Russian Military Power 1700–1800*, (London: Routledge, 2016), p. 236.
31. For example: Genrich Leer, *Voennaya strategiya* [Military strategy], (Moscow: URSS, 2019); Nikolai Mikhnevich, *Osnovy strategii* [The foundations of strategy], (Moscow: URSS, 2016); Evgeny Martynov, *Politika i strategiya* [Politics and strategy], (Moscow: 'Finansovyi Kontrol', 2003).
32. Vladislav Iminov, 'K 185-letiyu so dnya rozhdeniya G.A. Leyera' [On the 185th anniversary of the birth of G.A. Leer], Ministry of Defence of the Russian Federation, http://энциклопедия.минобороны.рф/encyclopedia/history/more.htm?id=11914040@cmsArticle (accessed 20 July 2020).
33. Leer, *Opyt kritiko-istoricheskogo issledovaniya*, pp. 21–22.
34. Kokoshin, *Soviet Strategic Thought, 1917–91*, p. 20.
35. For example: Thomas Rid, *Active Measures: The Secret History of Disinformation and Political Warfare* (London: Profile Books, 2020); Mark Galeotti, *Russian Political War: Moving beyond the Hybrid* (London: Routledge, 2019).
36. Martynov, *Obyazannosti politiki*, p. 20.
37. See the following discussion about the influence of Evgeny Messner.
38. See Ofer Fridman, *Russian 'Hybrid Warfare': Resurgence and Politicisation* (London: Hurst, 2018).
39. For example: Oscar Jonsson, *The Russian Understanding of War: Blurring the Lines between War and Peace* (Washington: Georgetown University Press, 2019); Galeotti, *Russian Political War*.
40. Leer, *Opyt kritiko-istoricheskogo issledovaniya zakonov vedeniya voyny (polozhitel'naya strategiya)*, p. 14. Italic emphasis is Leer's.
41. Evgeny Martynov, *Strategiya v epokhu Napoleona i v nashi vremiya* [Strategy in the age of Napoleon and in our times], (St Petersburg: Voennaya Tipografiya, 1894), p. 1. Italic emphasis is Martynov's.
42. Nikolai Mikhnevich, *Strategiya* [Strategy], vol. 2 (St Petersburg: Tipografiya Trenke i Fyusno, 1910), p. 36.

EDITOR'S INTRODUCTION

43. Leer, *Opyt kritiko-istoricheskogo issledovaniya zakonov vedeniya voyny (polozhitel'naya strategiya)*, p. 7.
44. Medem, *Obozrenie izvestniyshikh pravil i system strategiy*, p. 182.
45. It is important to note that in their writings on strategic combinations, Russian Imperial strategists commonly referred to the 1845 *De l'esprit des institutions militaires* by Auguste de Marmont, whose influence in this respect on the Russian understanding of strategy seems to be stronger than that of Clausewitz or Jomini.
46. Leer, *Opyt kritiko-istoricheskogo issledovaniya zakonov vedeniya voyny (polozhitel'naya strategiya)*, pp. 18–24.
47. Anton Denikin, *Ocherki russkoy smuty* [The Russian turmoil], vols. 1 and 2 (Paris: Izdatel'stvo Povolotzkogo, 1921–22), vols. 3 and 4 (Berlin: Slovo, 1924–25), vol. 5 (Berlin: Mednyy Vsadnik, 1926).
48. For example: Aleksander Savinkin and Igor Domnin (eds.), *Gosudarstvennaya oborona Rossii: Imperativy russkoy voennoy klassiki* [National defence of Russia: The imperatives of Russian military classics], (Moscow: Voyennyy Universitet, Russkiy Put', 2002); Aleksander Savinkin and Igor Domnin (eds.), *Zashchita otechestva: Nauka pobezhdat', zavety i uroki Petra Velikogo* [Defence of the fatherland: The science of victory, covenants and lessons of Peter the Great], (Moscow: Voyennyy Universitet, Russkiy Put', 2010); Alexander Savinkin (ed.), *Russkoye zarubezh'ye: Gosudarstvenno-patrioticheskaya i voyennaya mysl* [Russian émigrés: Political-patriotic and military thought], (Moscow: Gumanitarnaya Akademiya Vooruzhennykh Sil, 1994); Alexander Savinkin (ed.), *Voyennaya mysl' v izgnanii: Tvorchestvo russkoy voyennoy emigratsii* [Military thought in exile: The oeuvre of Russian military émigrés], (Moscow: Voyennyy Universitet, Russkiy Put', 1999; Aleksander Savinkin (ed.), *Ofitserskiy korpus russkoy armii: Opyt samopoznaniya* [The officer corps of the Russian Army: The experience of self-knowledge], (Moscow: Voennyy Universitet, Russiki Put', 2000); Evgeny Messner, *Vsemirnaya myatezhvoyna* [Worldwide subversion war], (Moscow: Zhukovskoye Pole, 2004); Andrey Snesarev and Anton Kersnovski, *Filosofiya voyny* [Philosophy of war], (Moscow: Veche, 2018); Nikolai Golovin, *Voennye usiliya Rossii v Mirovoy voyne* [Russia's military efforts in the World War], (Moscow: Kuchkovo Pole, 2001).
49. Alexander Savinkin, 'Zashchita Rossii' [The protection of Russia], in Savinkin, *Russkoye zarubezh'ye*, p. 8.
50. Igor Domnin, 'Kratkiy ocherk voyennoy mysli Russkogo zarubezh'ya' [A short outline of the military thought of Russian émigrés], in Savinkin, *Voyennaya mysl' v izgnanii*, pp. 523–24.
51. Boris Gerua quoted in Domnin, 'Gorenie dukha', p. 214.
52. Anton Kersnovski, *Filosofiya voyny*, p. 13.

STRATEGIYA

53. Kokoshin, *Politologiya i sotsiologia*, p. 170.
54. Nikolai Golovin, *Nauka o voyne: O sotsiologicheskom izuchenii voyny* [The science of war: On the sociological research into war], (Paris: Izdatel'stvo gazety 'Signal', 1938), p. 28.
55. Leer, *Opyt kritiko-istoricheskogo issledovaniya*, p. 19.
56. Golovin, *Nauka o voyne*, p. 158.
57. Ibid., p. 22.
58. Ibid.
59. Ibid., p. 48.
60. See Fridman, *Russian 'Hybrid Warfare'*.
61. Igor Domnin, 'Zaklyucheniye: "Dusha armii"—vzglyady Russkoy voyennoy emigratsii' [Conclusion: 'The soul of the army'—The views of Russian military émigrés], in Igor Domnin (ed.), *Dusha armii: Russkaya voyennaya emigratsiyao moral'no-psikhologicheskikh osnovakh rossiyskoy vooruzhennoy sily* [The soul of the army: Russian military émigrés on the moral and psychological foundations of the Russian armed forces], (Moscow: Voyennyy Universitet, Nezavisimyy Voyenno-Nauchnyy Tsentr 'Otechestvo i Voin', Russkiy Put', 1999), p. 286.
62. See Fridman, *Russian 'Hybrid Warfare'*.
63. Evgeny Messner, *Lik sovremennoy voyny* [The face of contemporary war], (Buenos Aires: South American Division of the Institute for the Study of the Problems of War and Peace named after Prof. General N.N. Golovin, 1959), p. 5.
64. James C. Pearce, *The Use of History in Putin's Russia* (Delaware: Vernon Press, 2020), p. xv.
65. Savinkin, 'Predislovie: Zashchita Rossii', p. 8.
66. Igor Domnin, 'Vvedeniye: Dukhovnyy smysl Russkoy voyennoy emigratsii' [Introduction: The spiritual value of Russian military émigrés], in Domnin, *Dusha armii*, p. 9.
67. Editors, 'Introduction', in Savinkin, *Ofitserskiy korpus russkoy armii*, p. 9.
68. For example: Dima Adamsky, 'Cross-Domain Coercion: The Current Russian Art of Strategy', *Proliferation Papers*, no. 54, November 2015; Timothy Thomas, 'The Evolution of Russian Military Thought: Integrating Hybrid, New-Generation, and New-Type Thinking', *Journal of Slavic Military Studies*, vol. 29, no. 4, 2016, pp. 554–75.
69. George Orwell, 'Politics and the English Language', *Horizon*, vol. 13, no. 76, pp. 252–65.

1

THE EXPERIENCE OF HISTORICAL-CRITICAL RESEARCH INTO THE LAWS OF MILITARY ART (POSITIVE STRATEGY)[1]

Genrikh Antonovich Leer

I

The value of war among other different phenomena in the lives of societies. War as one of the most rapid and most powerful civilisers of societies.

War is one of the instruments in the hands of politics, the most extreme instrument to achieve a state's goals. It is a dispute over rights between states, which are considered political powers.

Frequently war is seen as an exceptional evil, and military power as an extremely burdensome and unproductive power of the state. Those

---

[1] This chapter is a translation of the 'Introduction' from Genrikh Leer, *Opyt kritiko-istoricheskogo issledovaniya zakonov vedeniya voyny (polozhitel'naya strategiya)* [The experience of historical-critical research into the laws of military art (positive strategy)], (St Petersburg: Pechatnya V. Golovina, 1869).

who see war in this light are either incapable of seeing the whole picture from all its sides or do not have the composure required to analyse the problem. The main cause of all these wrong conclusions is human nature in general, which is always one-sided and incapable of the composure required to properly analyse the problem in question. War can be seen as pure evil only by those who examine it from a very narrow angle. Those who isolate from the general history of humankind only the periods of war will, indeed, see only bloodshed, destruction, and the deaths of individuals and whole societies.

However, if one takes a wider perspective that includes the period before a war as well as that after it, one would see a completely different picture. Even a quick glance at this wider picture will immediately allow one to understand the enormous step forward that a society involved in war makes in its internal development. This would help one to comprehend the series of fundamental reforms that usually follow every war—reforms that nobody would think about otherwise or whose development would involve a significantly longer period of time without this powerful impulse called war.

History is full of examples that support this understanding. For example, it is worth recalling Prussia before and after the 1807 Treaties of Tilsit or Austria before and after the 1866 Austro-Prussian War. A close analysis of these examples will convince anyone that war is a powerful engine that pushes the domestic moral and material affairs of societies forward. After such an analysis, anyone will realise that destruction and killing form only one side of war; the other side shows that *war is one of the most rapid and most powerful civilisers of societies.*

Therefore, according to Plutarch's accounts, the bloody wars of Alexander the Great were of great service to undeveloped Asia, by giving it the Greek language, Greek arts and Greek civilisation. These wars, by the way, introduced Asia to legal marriage, one of the main foundations of civilised societies, involved the building of seventy cities, etc.

There is no doubt that on the way to a better civilisation and the improvement of material prosperity it would be very desirable to avoid war. It is important to sympathise with this desire and not to lose hope that *one day, in the distant future, this dream will become true*—those who deny this deny the laws of continuous human progress. However, analysis of human history, combined with an assessment of the state of contemporary affairs, convinces us that we are far from the moment when humanity will find it possible to reject war. It is worth recalling the origins of many European states and the history of their formation (fortunate marriages, successful wars, etc., entered into solely for the purpose of territorial enlargement), to remind ourselves that in this transitional period of ours, wars will occur very frequently.

As discussed above, despite its evil side, war has also productive outcomes. It is important to note, however, that history is full of examples of wars that had no productive consequences whatsoever. Both sides of such wars are evil, like the 1701–14 War of the Spanish Succession, which had not even a minor positive impact on the lives of people. Unfortunately war can be abused. While any known means can be abused to produce evil, it should not automatically lead one to the conclusion that such means are an absolute evil and have no other side, including a good one.

To conclude, *war is a quite natural phenomenon in the lives of societies*, as struggle is an inherent part of everything living. *While war has quite a large evil side, when it is used judiciously it is one of the most rapid and powerful civilisers of humanity.*

II

Military Art. Its purpose. It is more difficult and more complex than other arts. The qualities required by war from a military person.

History shows that small armies sometimes beat an enemy that is significantly more powerful (in *the Battle of Marathon*, 10,000 Greeks beat 100,000 Persians; in *the Battle of Alesia*, 40,000 Romans beat 280,000 Gauls, etc.). An explanation for this phenomenon can be found in the art of using military forces. The skilful deployment of forces or, in other words, the proper and skilful direction of forces in the theatre of military action (as well as on the battlefield) to achieve the aims of war in the shortest time and with the least sacrifices—this is what constitutes the main goal of military art.

Military art is more difficult and more complex than other arts, because in these (painting, sculpting, etc.) the masters work with dead elements, which can be measured and weighed, and which passively submit themselves to external influences. Military art, however, in addition to other elements, includes *the human being as the main instrument of war* (the unchanging nature of the human being's physical characteristics, combined with the infinite variety of his moral characteristics, makes it very difficult to control him in general, and especially during the critical moments of combat).

Another quality of military art that separates it from other arts is the fundamental importance of *time* (*Napoleon* used to say, 'achievements in war depend on *glazomer*[2] and timing' and 'in war and politics, a moment once lost will never return'). While masters of other arts create only when they feel inspired, during a war one has to be inspired at every given moment (*like Napoleon's bravery which has to be found at two after midnight*).

Finally, while painters, sculptors, architects, etc., combine the *dead* elements of their arts into one whole product *in a tranquil situation*, the masters of military art combine *live* elements *in nerve-racking situations* that can unsettle even people with the most powerful natures.

All these reasons explain best why great military leaders, the great masters of the art of war, are very rare. An analysis of the qualities required by war of a military person provides additional reasons for this rarity.

Resourcefulness—an ability to find solutions in any possible scenario—is the most fundamental quality required of a military person. It consists of *mind*—an ability to discern the best way of action according to known conditions—and *character*—an ability to decide on the execution of this way in the shortest possible time. There is no doubt that both mind and character are equally important for military commanders. However, if one has to make a choice, then the first place should be given to character.

Therefore, *military affairs are predominantly the affairs of character*. As Clausewitz rightly states: 'in war, what is important is not to dare to do the best thing, but *to dare to do anything, as far as this anything can be energetically executed*'. 'The worst thing that one can decide in war', writes

---

[2] The Russian word *glazomer* (*coup d'œil*—literally, an ability to measure distances by eye) describes the ability to assess a situation.

Jean-Thomas Rocquancourt,³ 'is to decide nothing.' 'Indecisiveness in war', states Michel Ney,⁴ 'is the most fundamental weakness of a leader, especially at a time when the enemy is approaching. Instead of wasting time in vain on lengthy discussions, it is vital to take *any* decision.'

'All the qualities required of a commander', states Napoleon, 'are very rarely combined together. It would be most desirable that the *mind and talent* of this person be balanced *with his character and bravery*, a quality that already distinguishes such a person from the general mass of people. He should, metaphorically speaking, be like a square, with the base (character) equal to the height (mind) ... If bravery prevails in a general, then he will make the mistake of daring to engage in actions which are beyond his abilities. And on the other hand, he will never dare execute his decisions if his character and bravery are less than his mind.' 'There is nothing more difficult and, thus, more important', stresses Napoleon, 'than the ability to take decisions' because 'the true wisdom of a general is expressed in his ability to take energetic decisions'.

'The *primary* quality of a general', states Maurice de Saxe,⁵ 'is his character (bravery). Without it, I do not consider the others, because they become useless. The *second* is mind ... and the *third* is health.'

'Two things', argues Auguste de Marmont,⁶ 'a general must, therefore, possess: *mind* and *character*. Mind—because, without it, no combinations can be made; one surrenders without any defence. Character—

---

³ Jean-Thomas Rocquancourt was a French military commander, a professor of military art and military history at Saint-Cyr Military Academy (1821–46).
⁴ Michel Ney was a French military commander and Marshal of the Empire who fought in the French Revolutionary Wars and the Napoleonic Wars. He was one of the original eighteen Marshals of the Empire created by Napoleon I.
⁵ Maurice, Count of Saxony was a famed military commander of the 18th century. The son of Augustus II the Strong, King of Poland and Elector of Saxony, he initially served in the army of the Holy Roman Empire, then the Imperial Army before becoming a Marshal General of France.
⁶ Auguste de Marmont was a French general who rose to the rank of Marshal of the Empire. In the Peninsular War, he succeeded the disgraced Andre Masséna as commander of the French Army in northern Spain.

because, without a strong and obstinate will, the execution of conceived plans will never be secured. But here the relative qualities prevail over the absolute qualities, and character should control mind. In this state we find the element of success. If every quality could be expressed in figures, I would prefer a general having five parts of mind and ten parts of character to one who has fifteen parts of mind and eight parts of character. Whenever character has the ascendancy over mind, and the latter is of a certain extent, the chances are that an object determined upon will be attained. But if mind is superior to character, the judgement, projects and the direction are being continually changed, because a man of vast intelligence considers all questions at every instant in a new aspect. If strength of will does not interpose between these continual changes, we float for ever between two parts, irresolute which one to take. We end by choosing none whatever (which is worse still); and, instead of approaching our object, an uncertain march brings us further from it, and we are lost in a wilderness.' ('Esprit des institutions militaires', pp. 264 and 265.)[7] {An example of the ill consequences caused by vast intelligence combined with lack of character can be found in the actions of Archduke Charles, Duke of Teschen at the beginning of the war of 1809.}

In addition to all the qualities already mentioned that are required of a military person, it is important to remember another, quite significant one—*glazomer*, which is *a continuous accurate assessment of the situation, time, conditions and space required to achieve one or other combination*. It is obvious that *glazomer* (in other words *Military Diagnostics*) plays a very important role in a matter that depends on the ability to act according to the situation, which is in a state of constant flux and is liable to rapid and infinitely variable changes.

We have tried, by means of analysis, to highlight the qualities required of the masters of military art. Even separately, these qualities—*character*, *mind* and *glazomer*—can be found very rarely, not to mention at the level required by war. History is the best proof of the

---

[7] Leer refers to Auguste de Marmont, *De l'esprit des institutions militaires* (Paris: J. Dumaine, 1845). The quote is from Auguste de Marmont, *The Spirit of Military Institutions*, translated by Frank Schaller (Columbia: Evans and Cogswell, 1864), pp. 229–30.

rareness with which these qualities are fortunately combined to create a military genius. In the last twenty centuries it happened no more than ten times: Alexander the Great, Hannibal, Julius Caesar, Gustavus Adolphus, Turenne, Prince Eugene of Savoy, Peter the Great, Frederick the Great, Alexander Suvorov and Napoleon. Nature is very ungenerous in creating true masters of the arts, and it seems that in military art it is even less generous, because this art is more difficult and complex than the others.

Obviously other arts also require similar qualities, but do they require them on the same scale and in the same form? For example, other arts also require an eye, but usually a geometric eye. Other arts also require character. But it is not the same character required in the critical moment of an awful situation to compel tens of thousands of soldiers to voluntarily face mortal danger. Other arts also require mind. But it is not the same mind that can be inspired in every given moment, because even a brief delay in inspiration during war can ruin the whole affair. In other arts, mind does not depend on time: masters can toil on their creations for years or even decades, working on them only when they feel inspired. In their case the inspiration is never late, and every mistake can be fixed.

III

Military art, like any other art, is based on unchanging laws, whose application varies infinitely depending on the constantly changing environment. The meaning of law (theoretical rule). Science (a theory of art) and art. Neglect of the study of military art. In discussing any science, conceptualisation[8] and history should work together.

There is no doubt that military art, like any other art, is based on unchanging laws. It is true because laws form the foundation of all our activities, starting with eating, walking, etc., and extending to the highest forms of political and strategic thinking. In the former, as well as the latter, any violations of law are ruthlessly punished. Regarding

---

[8] Leer uses the word 'philosophy'; however, in a modern context, what he meant was 'conceptualisation'.

eating, for example, it is known from experience that *a human needs to eat two to three pounds per day to support his existence*. Anyone who tries to violate this law in either direction would find that in both the final result is the collapse of the body. From experience too, it is known that *a human can walk at the speed of four kilometres per hour*. Anyone who follows this law can walk effortlessly up to twenty kilometres per day if he alternates walking with resting. If someone tries to violate this law and do six kilometres in the first hour, then in the second he will barely make two kilometres, and in the third he will stop entirely.

When we move to the highest spheres of human activity, such as politics, war, etc., we find similar laws. For example, *war cannot serve as an end in itself, but is merely one of the means, and quite an extreme one, in the hands of politics to achieve a state's aims*. This is a law. Any attempt to violate this law and turn war, as Napoleon used to do, into an end, and anything else, including politics, into the means of war, will result in numerous and immense political mistakes, similar to those that led France from its ephemeral greatness to the edge of destruction, bringing an end to the political life of the person responsible for these mistakes. {An example of turning war into an end can be seen in Napoleon's preparations for invading England and his extremely forceful extraction of money from his allies for this purpose.}

Another law related to the world of politics and war is that *the military establishment of a state should correspond precisely with this state's capabilities and means, as well as with its wealth*. A violation of this law, similar to the violation of any law, leads to harmful consequences. If the military establishment is smaller than other capabilities and means of the state, it will weaken or even destroy its *external* political status. The opposite situation would also be harmful, leading to the state's *internal* dissolution.

Focusing our analysis directly on military issues, we can point to laws relevant to every issue in any situation, providing examples that prove that a violation of this law will have harmful consequences. Let's take the example of military uniform, which seems to be (but is not) one of the more minor issues. Theory says: '*clothe a soldier in a way that he will be protected in all possible weather conditions, as well as feel comfortable in battle. Make the clothes ordinary, allowing some external attributes to make it easier to muster and observe discipline. Following all these requirements, ensure that the uniform is cheap and, please, smart.'* To put it briefly—the law puts the requirement of *comfort* first and the requirement of *smartness* last. Having this *law or rule* in mind, let's analyse how the issue of uniform was solved at different times in the past. For example, let's analyse Prussian uniforms of the 17th century. The whole of Europe decided to wear these uniforms, never mind that they were ugly— though this is a question of taste, as at the time the uniform was considered handsome. What was the price paid for this handsomeness? According to *Henry Lloyd*,[9] the *problems* related to this uniform sent, initially to hospital and later to their graves, a fifth to a quarter of all armies in every campaign (!). So what? Even though the experience of war mercilessly and visibly punished soldiers for such criminal violation of the law, the ugly Prussian uniform (against which our great Suvorov argued so much) existed almost for half a century. If the estimate made by Lloyd is right, and if it was possible to add together a fifth to a quarter of the soldiers deployed by all armies during this period, then we will arrive at an enormous number of innocent and futile victims of the capricious violation of the law described above, i.e., putting the uniform's appearance ahead of other requirements.

A sensible theory argues: *projectile weapons (firearms) are preparatory weapons, and cold weapons are decisive weapons.* {It is important to point to very original formulation of this law given by our great general Suvorov in his famous formula: 'The bullet is a fool, the bayonet a fine chap.'} This is also a law which, regardless of different technological improvements, defines fixedly the relations between these two types of weapons. {This law is best understood by those who make different claims after each minor improvement in firearms!} And again, a violation of this law, like the violation of any law, leads to harmful consequences. It is worth recalling the consequences created by the desire, which was especially strong in the 16th century, to give firearms the role of decisive weapons. This created a situation

---

[9] Leer refers to Henry Humphrey Evans Lloyd, who was a Welsh army officer and military writer. He fought for the French against the Austrians, the Jacobite forces of Charles Stuart against the British, the Austrians against the Prussians, and the Prussians against the Austrians (during the same war), and the Russians against the Turks.

where battles were not battles at all, but duels fought with firearms. The immediate consequence of this was that battles lost their significance as the most decisive tactical means, leading to decades-long wars without any results.

Furthermore, a sensible theory states: '*one will usually beat one's enemy if at the decisive time and the decisive place one is stronger than the enemy.*' Again, this is a law. An examination of all wars and battles through the prism of this *law or rule* will show that success lay always with those who respected this law, and failure awaited all who violated it. Let's take for example the 1866 Battle of Custoza, where the Austrian military beat the twice-larger Italian army. The main reason for the Italian defeat was that the Italians violated this law. As is well known, Alfonso La Marmora sent into battle only seven of his twenty divisions. Moreover, five forward-facing divisions were dispersed across ten kilometres, creating a large dispersion instead of a concentration of force in a decisive place. This major mistake led to many others, such as the deployment of divided units, etc.

If we examine the question of *choosing positions*, then there is again a law (*rule*) that defines such decisions: a known combination of conditions, which any position has to meet. Here, too, a violation of this law, like the violation of any law, leads to harmful consequences, and the best example of it was the *1807 Battle of Friedland*.

Furthermore, theory offers another law regarding the question of *operational lines*: an operational line *has* to fulfil the following requirements: (1) it *has* to lead towards an important goal; (2) it *has* to be shortest and most convenient possible; and (3) it *has* to be safe. {It is impossible to leave unnoticed the fact that the very language of the theory expresses the power of the law.} The consequences of following this law are best demonstrated by Napoleon's *1805 Ulm Campaign*. The consequences of violating this law are best demonstrated by the manoeuvres of the Allies before *the 1805 Battle of Austerlitz*. The Allies' operational line did not fulfil the first requirement, because Napoleon, after skilfully changing his operational line to Jihlava—Regensburg, saw no importance in his line to Vienna. It also did not meet the second requirement, because a direct march from Vyškov was only fifteen kilometres, but the indirect strategic detour took three days. Finally, the third requirement was not fulfilled either, because the dangerous detour exposed the Allies' flank to the enemy, which was only fifteen to twenty kilometres away. The consequences of these decisions are known to all.

Generally speaking, for any issue in the field of military art, theory offers us a law, *a theoretical rule* or, as Napoleon stated, *the axes to allow the drawing of curves*, respect for which leads to success, while violation leads to harmful consequences. {There are also other laws that can be listed here, such us the unification of power in the hands of the general commander; allowing him full freedom of action; the extreme exertion of all capabilities and means to achieve the desired aim, etc. An analysis of these laws and their roles in the art of war lies at the heart of this work. For now, however, we merely give some examples to demonstrate that military art, like other arts, has its own unchanging laws.}

The number of laws that serve as the basis of military art is not big. They are unchanging for all time, but their application varies infinitely *depending on the situation*. In mathematical language, all these general laws are formulas that always lead to the same outcome, i.e., regardless of any possible way of solving the problem, they embody the correct notion of a variety of possible solutions.

Using the brilliant perspective of Peter the Great, these laws define 'only the general course, and not actions for the specific *time* and *occasion* (i.e., situation)'. In other words, there are no laws (*rules*) that suit every possible occasion, because the number of possible occasions is infinite, as, according to Napoleon, 'all questions of higher tactics seem like indeterminate physical-mathematical problems that allow multiple solutions'. Therefore, each law should be seen as a formula in which the situation (*time* and *occasion*) is a variable that should be inserted to arrive at a specific solution to every known occasion. {Moreover, as in mathematics, possible solutions can be both positive and negative.} To put it straightforwardly, a law should be applied according to its *spirit*, and not *slavishly* according to its letter, as, in the words of Napoleon, 'in all military sciences, theory is useful to give general ideas that form thinking; its slavish implementation, however, is always very *dangerous* ...'

Every law can be formulated either theoretically *a priori*, or practically *a posteriori*. The laws of military art have been mainly formulated in the latter way, i.e., *by observing real activities and specific occurrences, and abstracting their general characteristics.* {As, according to Napoleon, 'the true origins (laws) of war (military art) are those that were followed by great military leaders: Alexander, Hannibal ... whose great deeds are brought to us by *history*.'} The way in which a law is formulated does not really matter; what matters is that it is a law. [...]

A corpus of theoretical laws, systematically assembled by elucidation and analysis based on an investigation of the characteristics of the elements of the laws and the relations between them, constitutes *science* (or, in fact, *a theory of art*). The implementation of these laws in the infinitely varied conditions of life is, in fact, a matter of aptitude, which is separate from and above science, as it is *an art*. In this way, when it comes to the implementation of any science in life, it turns into art—geometry turns into surveying, pathology and anatomy into medical art, mechanics into the art of building machines and other different constructions.

Even though military art is more difficult and complex than other types of art, and is built on unchanging laws, it does not receive the attention it deserves (a fact known at least from the times of Lloyd). The immediate consequence of this unfortunate state of affairs is that there are no other arts in which one can find so many dilettantes. {Everyone thinks that he has a right to have his own opinion, regardless of how ridiculous it sounds; such as various claims that in war everything depends on chance [...] Especially emphatic verdicts are given by those who 'study nothing, but know everything'.} There are no other arts in which caprice, fashion, routine and the spirit of blind imitation have as deep roots as in military art.

Who should take the blame for these unfortunate circumstances? Should it be the general populace who have remained indifferent to studying this art? Partially yes. The major part of the blame, however, should be put on the science itself, which has failed so far to attract the general population and acquire their trust.

There are several reasons for this:

1. Science has been falsely interpreting the meaning of laws. Laws, by their very nature, are general formulas, which should be adjusted each time according to the situation (*time* and *occasion*). Thus, there simply cannot be laws (so-called *rules*) that apply equally in every situation, because these situations are infinitely different. Unable to

grasp this nature of laws, science has been striving to turn them into *rules or prescriptions*.
2. Science has been basing itself either on general theoretical and conceptual conclusions, unsupported by sufficient observations and facts, or on pure historical observation of facts without significant analysis and generalisation. In other words, science has been taking either conceptual or exclusively historical approaches, trying to distinguish between them while, in fact, they complement one another.

There is only one way to eliminate this misfortune: to bring these two different approaches together, because one complements the other. *Conceptualisation and history* (a critical-historical way of analysis, see paragraph VI below), according to Professor Johann Kaspar Bluntschli,[10] *inevitably have to combine in their attempt to analyse any type of science.* The main problem is that until today, *theory* (science) and *history* (the application of science to life) have been going separate ways, if not one against the other. Hence, theory or science has frequently turned into scholasticism and metaphysics, and history into a mass of insufficiently explained plain facts that clog our memory.

IV

The theory of military art (science). Its purpose. The elements of military art and the military sciences of these elements. The relations between the elements and sciences of these elements. Strategy as a synthesis of military affairs.

Assembling and elucidating laws that constitute military art, investigating the characteristics of their elements and the relations between them in infinitely changing situations—all these constitute the main aim of *the theory of military art*, i.e., *military science*.

The elements of military art can be divided between: (I) the *moral* element; and (II) the *material* elements.

---

[10] Johann Kaspar Bluntschli (1808–81) was a Swiss jurist and politician who developed one of the first codes of international law and war.

STRATEGIYA

I. Moral Element

In a direct sense, the moral element forms the moral strength of an army. In a more general sense, it forms the moral strength of the whole nation (the *heart* and *mind* of the army and the nation, because they both participate in war).

The moral element plays a very important role in war, as, according to Napoleon, three-quarters of success in war depends on it, and only a quarter can be attributed to various material elements. Since it is impossible to calculate or measure the moral element, it is very difficult to analyse it theoretically. However, it would be a mistake to reject the possibility that, in the distant future, newly discovered laws will define the ways by which hearts and minds function in war. Thus, the possibility of *military psychology* should not be rejected. {*What seems to be an illusion or utopia today can become a fact tomorrow*. The truth of this claim has been supported by history on too many occasions.} However, the only thing that can be done nowadays is to indicate the right way that leads towards the future discovery of these laws. As with any other art, the way to discover new laws should be a *way of analysis* that is based on observing individual occasions (in our case, facts provided by military history) and the subsequent extraction of common (to all occasions, or at least to a majority of them) characteristics, i.e., *a move from analysis to synthesis*. In this way, based on the common characteristics extracted, it will be possible to suggest *a general conclusion* or *a law*. {All laws of chemistry have been discovered in this way [...], in artillery it helped to discover the laws of trajectory. (Already it is possible to recognise the great importance of *critical military history* to a military man, as for him it is like a *laboratory* for a chemist, an *observatory* for an astronomer, a *shooting range* for an artillerist.)} Therefore, while military psychology is a science of the distant future, it is still important to pursue this field now. It is important to observe individual occasions and analyse the most significant facts related to the moral element in war—creating *military-psychological records*, until the time when *military psychology* arrives to study them.

II. Material Elements

1) *The tactical element on the battlefield* (in the army and the navy) (*tactics* and *artillery*) is the first of the material elements, because *the human*

being is the major weapon in war. This element is analysed by tactics. Its purpose is to solve the questions *about the most advantageous deployment of forces in the different situations of war, as well as their most advantageous organisation according to the defined goal*.

Weapons can also be included within the tactical element, as additional means that complement humans' natural means for attacking their adversaries. Questions about the most advantageous arrangement of weapons in accordance with the different situations of war, as well as their most advantageous deployment, are analysed by *artillery science*. In analysing the latter question, tactics also plays a part.

2) *The element of terrain* (*military geography*, *military topography* and *fortification*) is the next element. A war or a battle is waged on a certain terrain, and thus this element is included in all military calculations. In the plan for a campaign it is included in its most general sense—as *a geographical element*. It defines large and wide spaces restricted by large natural boundaries—*the theatres of military action*. In battle plans, the element of terrain is taken into consideration in its narrower sense as *a topographical element* and its role on *the battlefield*.

Terrain can either help or hinder the ability of military forces to achieve their defined goals. To assess the degree to which the terrain helps or hinders the ability of military forces to achieve their defined goals, it is necessary to draw up a plan of the terrain or, at least, to assess the characteristics of the terrain using a pre-existing plan. An analysis of these tasks constitutes the main purpose of *military topography*.

If a specific terrain hinders the ability of military forces to achieve their defined goals, then the means must be devised to eliminate these unfavourable conditions. It is important to prepare an advantageous theatre of war, as well as an advantageous battlefield—this is the main task of the *art of military engineering* or *fortification*.

3) *The administrative element* (*military administration*) is the next element. *Tactics*, as stated above, is responsible for questions about the most advantageous deployment of forces in battle. However, before this can happen, it is necessary to discover the means that offer the most rational solutions to other various questions related

to the most advantageous ways to equip, organise and supply forces with everything required during peacetime and war. All these questions constitute the main purpose of *military administration*.

4) The level of development of all these elements as a whole defines the political significance of the state, as well as its relations with its neighbours. This leads to a new element—the *political element* (*internal* and *external*—the former includes the influence of forms of governance on the formation of military forces, as well as on the form of waging wars). This element has a very important influence on military calculations.

An analysis of the influence of political conditions on waging war or, to put it simply, the link between war and politics should constitute the field of a separate though still non-existent military science—*military politics*. This science would be of great importance, because it would create and define a very close relationship between all military sciences that together constitute one large branch of science and of the *social* (*political*) sciences.

5) The next element is *military statistics*, the purpose of which is to assess the level of power of any known state at any given moment, i.e., its presence in all the elements discussed above. *Military statistics* are, in fact, nothing more than general statistics adapted to military goals.

6) *The element of chance* is an additional important element that can be both *external* and *internal* (the latter should be understood as the result of the capricious intervention of the human heart in affairs of war). Any plans of war and their execution are commonly surrounded and influenced by chance, unforeseen even by the brightest and best-provided minds (the 1757 Battle of Kolín, the 1702 Battle of Cremona, the 1809 Battle of Aspern-Essling, the 1800 Siege of Fort Bard, the 1814 Battle of Soissons, etc.). This element is the main reason why mathematical calculations can never fully explain the matter of waging wars. Napoleon used to say that 'an enterprise is well considered when two-thirds of the prospects are based on calculations and a third on chance. The best advice to anyone who is ready to leave nothing for chance is to undertake nothing.' Discussing this topic, Frederick the Great similarly concludes: 'waging war requires art and *luck*. Sometimes *unfortunate*

events occur, predictable neither by the best prudence in the world nor by the most meticulously conducted calculations.' Since war is not a matter of strict mathematical calculations, it turns to a certain extent (depending on the talent of those who lead it) into a game or, more correctly, *waging war is a type of probability theory*.

Clearly, the element of chance cannot be theoreticality analysed. However, the best way to understand the degree of its influence on the matter of war is, without a doubt, military history.

There is a very close relationship between all the elements described above. *A war is waged on a certain terrain* (the terrain element) *by humans* (the tactical element) *gifted with volition and desires* (moral element) *and all the needs of biological beings* (the administrative element). Finally, *a war is waged not for itself, but for achieving certain political benefits* (political element).

The characteristics of all these elements are studied by the military sciences. The interrelationship between these elements and their combination for the successful achievement of the goal of war in the specific context of a given time and space—all these constitute *the synthesis of all military affairs, i.e., strategy*. Therefore, the goal of strategy is *to grasp the question of waging war at a given moment in all its aspects and solve it according to the prevailing situation, i.e., to define a reasonable goal and direct all forces and means towards its achievement in the shortest time and with the least sacrifices*.

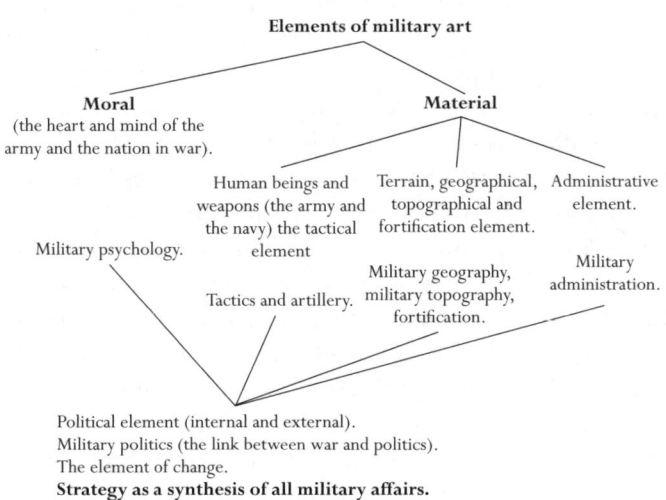

Political element (internal and external).
Military politics (the link between war and politics).
The element of change.
Strategy as a synthesis of all military affairs.

STRATEGIYA

V

For how long has strategy existed (as an art and as a science)? A brief review of the literature on strategy.

For how long has strategy existed? The answer to this question is probably found in history.

If we consider the main goal of strategy as *to define a reasonable goal and direct all forces and means towards its achievement in the shortest time and with the least sacrifices*, then we should see all military activities in which this goal was achieved as a manifestation of strategy as *an art*.

Strategy, as *an art*, has existed since antiquity. Moreover, its foundations have not changed since then. While the implementation of strategy and tactics infinitely changes, their principles have always remained unchanged.

Even a brief comparison between some similar events that took place in different periods of history, as distant one from another as possible, can easily prove this.

For example, it is possible to compare the initial actions of Alexander the Great during his landing in Anatolia in 334 BC and the landing of Gustavus Adolphus in 1630 near Peenemünde. By comparing these two similar operations which occurred in almost similar conditions, it is difficult not to see the great similarity in how the goals intended *to prepare the theatre of military action* were defined, as well as how they were achieved. {Taking into consideration one main difference which can be explained by the difference in the environment in which each of the generals found himself. The first, Alexander, immediately attacked the enemy army (the Battle of the Granicus) and started preparing the theatre of military action only after defeating it. The second, Gustavus Adolphus, evaded battle and immediately started preparing the theatre of military action.} After their initial landing, both generals found themselves in a very similar situation: both had sea behind them and both had a very narrow base on land. This led them to define the same goal and choose almost similar means to achieve it. Both sought to safeguard their rear and widen the land base by conquering the most important points on the coastline, thereby securing the lines of communication (in the first case with Greece, and in the second with Sweden). Only after achieving this goal, both turned to more decisive actions against their respective enemies.

GENRIKH LEER

It would be difficult to find two historical events that are so distant from each other in time, but so close in their contents as well as external appearance. It is very difficult not to see in them the appearance of strategy. In both cases we can clearly observe how both generals defined a reasonable goal and directed all available means towards its achievement in the shortest time and with the least sacrifices according to the requirements of their respective situations. Therefore, we can clearly recognise in these actions a manifestation of true strategy.

Another example is a comparison between the march of Hannibal through the seemingly impassable, overflowing Arno (217 BC) and the march of Napoleon from Valenza to Piacenza (the 1796 Battle of Fombio). Here too it is difficult not to notice that the successes of these two quite similar operations can easily be explained by the similarly fine choice of operational line and excellent execution of manoeuvres, which were *unexpected, unobserved and rapid* (characteristics that have always lain, and should always lie, at the heart of manoeuvres).

Finally, an additional example can be provided by two other events that were as similar in character as they were distant from one another in history: *the actions of the Romans after the Battle of Cannae and the actions of the Allies in 1813 immediately after the Truce of Pläswitz*. Here, again, it is difficult not to notice the similarity between the situations in which each side found itself, between the goals that each side defined for itself (both Romans and Allies aimed at *evading the battle*), as well as between the means to achieve these goals chosen by each side (*a fine combination of marches and protected camps*, which the Romans used to secure their overnight stays; and *a fine combination of manoeuvres and the Trachenberg Plan* in the case of the Allies).

These examples demonstrate that *the principles of strategy are immutable and that strategy has existed throughout all ages, as long as wars have been waged as an art*. This strategy, however, existed only in the minds of a number of gifted personalities, such as Alexander, Hannibal, Julius Caesar, Gustavus Adolphus, etc. This was so-called *natural or intuitive strategy*. It was intuitive because these great generals of the past, generously endowed by nature with characteristics required of a military commander, acted solely according to their genius and probably did not even assume the existence of certain deductions and laws that lay at the heart of their actions.

STRATEGIYA

In the middle of the 18th century, there was an attempt to make strategy, which previously had been the preserve of a small number of gifted personalities, accessible to the wider population. This was intended to make it accessible for purposes of understanding only, and not for its implementation, because the implementation of strategy will always remain the exclusive quality of only gifted and talented people. In other words, an attempt was made *to create a treatise on strategy*.

The literature on strategy begins with Henry Lloyd. {Originally an Englishman,[11] he served in the Austrian, Prussian and Russian armies. He was born in 1728 and died in 1783 in Holland in extreme poverty.} He created a very wide canvas for future writers to draw upon for strategy. Not only was Lloyd the first writer on strategy, he was the first to write a composite treatise on military affairs. While subsequent writers on strategy were mainly analysts, Lloyd did not restrict strategy by analysing only one side of military affairs or only one of the elements. Instead, he generously *pointed out* everything related to strategy. He did not focus exclusively on analysing military actions (the *tactical* element) {The first part of his work (De la composition des différentes armées anciennes et modernes) and the fourth (Des opérations de la guerre considérées en elles-mêmes)}; he continued further, giving his full attention to the most important element in war, which is generally excluded from military literature as something that cannot be calculated or measured, *the moral* element. {*The second part of his work (De la philosophie de la guerre)*}[12] Moreover, he went even further, analysing the *political* foundation of war when politics are contiguous with war. {In the third part of his work, Lloyd discussed the relations between different forms of governance and military actions.}

It would be very difficult to draw up a more complete and, at the same time, rational programme for the theory of strategy or, in other words, for strategy as a science.

---

[11] In fact, Henry Lloyd was Welsh.
[12] Leer refers to Henry Lloyd, *Reflections on the General Principles of War and on the Composition and Characters of the Different Armies in Europe*, Part II, chapter I, added to the second edition of his *The History of the Late War in Germany, between the King of Prussia and the Empress of Germany and Her Allies* (London: S. Hooper, 1781).

Indeed, as discussed above, is strategy not just a synthesis of all military affairs?

The essence of Lloyd's tactical advice (the first part of his work) can be summarised as follows: *respect the bayonet and consider firing weapons as preparation; shoot towards where you cannot move; move well, do not stop in the middle and finish the battle with a decisive blow.*

As a true writer on strategy, Lloyd was the first to establish clarity in the field of the most important question of strategy—*operational line*. In fact, he was the one who coined this term. On the one hand, when offering a theoretical rule (law) about operational line, Lloyd generally focused on how this issue can be resolved in contemporary terms (supplies of an army should come exclusively from supply depots; the five-march system;[13] *the tactical side of war is highly restricted by the administrative element*; an operational line, according to Lloyd, can extend to approximately fifty kilometres). On the other hand, he did not stop here but also pointed to another extreme form of operational line that can be *entirely independent of administrative restrictions*. Beginning his discussion on operational lines, he speaks about the Tatars' way of waging wars: 'The Tatars neither have, nor want depots; by the rapidity of their motions they must and do find everything on the spot. But when we penetrate, with our great and very heavy armies ...'[14]

Therefore, by pointing to these *two extreme* forms to resolve the issue of operational lines, Lloyd acknowledged the validity of all intermediate forms. {As, according to Napoleon, 'all questions of higher tactics seem like indeterminate physical-mathematical problems that allow multiple solutions'.} This also includes the contemporary form of operational line combining supply from depots with requisition. {The contemporary way of war, based on both supply depots and requisition (foraging), lies in the middle between the way of war in the 18th cen-

---

[13] The 'five-march system' is a military logistics concept that implies that an army should not proceed further from its supply depots than five marches at most. If it became necessary to proceed further, a fresh line of supply depots should be formed.

[14] The quote comes from Henry Lloyd, *Reflections on the General Principles of War and on the Composition and Characters of the Different Armies in Europe*, p. 133.

tury, based exclusively on supply depots, and the Tatars' way of war, based exclusively on extortion and foraging.} This leads us to conclude that Lloyd, in focusing primarily on strategy relevant to his contemporary environment, did not fall into a one-sided analysis that sought to systematise conditions relevant only to his times.

The most distinctive characteristics of Lloyd's work are concise and very precise views and discussions. Like all talented writers gifted with extraordinary creative minds, Lloyd neither meticulously described nor fully developed all his ideas. Instead, he restricted himself to drawing precise and creative strokes, leaving their development to his readers. In other words, Lloyd challenged his readers *to think and not to read*.

This is one of the main disadvantages of Lloyd's work. {This type of disadvantage was formulated best by Comte de Guibert,[15] who criticised Montesquieu's 'De l'esprit des lois', as well as Julius Caesar, Henri de Rohan, Raimondo Montecuccoli, Turenne, Maurice de Saxe, and Frederick the Great. 'The way of writing used by those great people', claimed Comte de Guibert, 'is marked neither by detailed development, nor by satisfying clarity. It seems that they were writing more for themselves, rather than attempting to educate the masses. This is a way always used by a genius who does not aim to educate. He always treats topics in the way that he sees them—quite casually, as he directs them from above. He does not descend to details. In this way, he bypasses intermediate ideas, required by an ordinary person to crawl with difficulty from one verity to another.' ('Discours sur l'état actuel de la politique, et de la science militaire en Europe', Chapter III.)}[16] This is the main reason why Lloyd's work has become inaccessible to the general masses, who *always prefer ready-made answers*. This is why many subsequent writers on strategy had to take upon themselves the painstaking work of the meticulous analysis avoided by Lloyd. Therefore, Lloyd was followed by a group of writers and analysts who discussed the issue of military actions either exclusively or primarily from the perspective of only one element. This group includes:

---

[15] Jacques-Antoine-Hippolyte, Comte de Guibert (1743–90) was a French general and military writer.

[16] Leer refers to Jacques-Antoine-Hippolyte, Comte de Guibert, *Discours sur l'état actuel de la politique, et de la science militaire en Europe* (Geneva, 1773).

1) *Dietrich Heinrich Freiherr von Bülow*[17] analysed military actions only from the perspective of the *administrative* element. {'Geist des neueren Kriegssystems' (1799) and 'Lehrsätze des neuern Krieges oder reine und angewandte Strategie' (1805)}[18] The main part of his analysis is concerned with lines of communication as well as the actions required for their creation and safeguarding. Another distinctive characteristic of Bülow's work is his effort to create a list of absolute rules. {Bülow acknowledged himself that his aim was to write a positive theory of military affairs, thus aiming to transfer them from the field of art (which requires a high level of creativity and can be implemented by only a few gifted persons) to the field of positive science (which can be studied by anyone). Therefore, he advises a defender *never* to stand in front of the attacker and *always* seek to stand on his flank, thus targeting his communication lines, leaving at the front a minimum force required to hold the enemy. An attack, according to him, should *always* develop in several concentric lines, and a retreat should *always* follow several eccentric directions, thus allowing for a shift from an eccentric retreat to a concentric attack.} Moreover, in an attempt to produce more concrete results, Bülow tended to illustrate his ideas by using geometric forms—a tendency that gives Bülow's work the character of so-called *strategic geometry*. {Bülow's work includes more than sixty geometric figures. Obviously, these forms make his ideas more concrete. On the one hand, the method of turning ideas into figures helps to comprehend them. On the other hand, if the same idea can take infinitely different forms in shapes depending on the infinitely changing situation, then it becomes clear that it should be treated not with *dead* geometric figures, but with the elastic and *live* images offered by military history.}

---

[17] Adam Dietrich Heinrich Freiherr von Bülow (1757–1807) was a Prussian soldier and military writer.

[18] Leer refers to Adam Dietrich Heinrich Freiherr von Bülow, *Geist des neuern Kriegssystems hergeleitet aus dem Grundsatze einer Basis der Operationen auch für Laien in der Kriegskunst faßlich vorgetragen von einem ehemaligen Preußischen Offizier* (Hamburg: Benjamin Gottlieb Hofmann, 1799); *Lehrsätze des neuern Krieges oder reine und angewandte Strategie aus dem Geist des neueren Kriegssystems* (Berlin: Frölich, 1805).

2) *Antoine-Henri Jomini*, whose first work ('Traité des grandes opérations militaires')[19] focused primarily on the *tactical* element of military action. In his view, the most important part of military affairs is the concentration of force, actions within internal lines, and battle at the decisive point (in complete opposition to Bülow). {Analysing the 1757 Battle of Leuthen, General Jomini was amazed by the way in which Frederick the Great, who had only 30,000 troops against an Austrian force of 80,000, succeeded in defeating the enemy in the most important and decisive moments of the battle. This analysis led Jomini to conclude that the essence of war is an ability to overwhelm the enemy at the most decisive point of the battle. This conclusion Jomini also applied to strategy, i.e., *in the theatre of military action*. Impressed by this important, though not sole, aspect of military affairs, Jomini primarily focused on it.}
3) *Archduke Charles, Duke of Teschen*[20] analysed the role of *the element of terrain* in military action. His work is based on the assumption that 'in strategic calculations, only terrain is a constant variable and all other elements can vary. Therefore, the decisive points in the theatre of war should be defined *solely* in terms of terrain, thus making other elements as definitive as the terrain itself.' Such a one-sided view of war could be conceived only in the Austrian military, which has always treated terrain with special respect (*defensive lines system*). {It is important to note that the significance of terrain can vary. Firstly, it depends on *the character of the state*. For example, in flat and cultivated states, road junctions turn out to be the most important military points; in mountainous areas, valley junctions take on the role of the most important points; in swampy areas, dry passages and bridges; in desert areas, oases and other places with access to water. Secondly, and most importantly, the significance of terrain changes according to *the place of an army in the theatre of military action*. Thus, using the words of General Jomini, one can distinguish between the *manoeuvre significance* of terrain and its *geographical*

---

[19] Leer refers to Antoine-Henri Jomini, *Traité des grandes opérations militaires*, vol. 1 (Paris: Chez Magimel, 1811).

[20] Archduke Charles Louis John Joseph Laurentius of Austria, Duke of Teschen (1771–1847) was an Austrian field marshal.

significance (i.e., the significance that this territorial place has *of its own regardless of military action*). While the geographical significance of a certain place is definite, its manoeuvre significance can vary infinitely. For example, the definite geographical significance of Piacenza is that control over it unites northern Italy (which is divided by the Po) into one theatre of military action. However, its manoeuvre significance from one military campaign to the next (1796 and 1800), as well as during these campaigns themselves, constantly changed according to the movements and relative positions of opposing armies in the theatre of military action. Another example is Verona in 1846. During the Battle of Santa Lucia, the fortified city seemed to be the most important position for the outnumbered forces of Joseph Radetzky von Radetz. However, as Radetzky managed to hold out until his reinforcements arrived, allowing him to shift the battles towards Milan and the Ticino River, Verona lost its status as the most significant territorial position and was turned into an operational base among many others. Quite often, territorial points with negligible geographical significance turn out to be the most significant places from the perspective of manoeuvre, like Polotsk and Vitebsk in 1812, or Champaubert in 1814. Therefore, it is difficult to agree with the argument that terrain represents a constant variable (though indeed it should constantly be part of any strategic calculations). Moreover, it is difficult to accept the proposition that the decisive points in the theatre of war should be defined *solely* on the basis of terrain. If this was true, then military actions in every theatre of war would follow the same scenario, and we all know that this is not the case.}

4) The works of *Joseph de Rogniat*[21] {Written under the influence of unsuccessful Napoleonic Wars in general, and the Peninsular War and the invasion of Russia in 1812 in particular.} share quite a similar approach with the works of Bülow: the important aspect of military affairs is occupied by lines of communication and their safeguarding. Choosing between two opposite principles of all military action—*decisiveness* and *prudence*—Rogniat almost entirely

---

[21] Joseph de Rogniat (1776–1840) was a French general of the Revolution and the Empire.

rejects the former. It is very strange to hear from this participant in the Napoleonic Wars that if an army moves more than eight days away from its original base, it should *stop* (!?) to set up a new one. The main purpose of the shift from supply depots towards requisition is to release military action from administrative shackles. Moreover, Napoleon's campaigns in 1805–9 serve as the best examples of brave and decisive military action without any preliminary preparation of the theatre of war from an administrative perspective (excluding the preparations conducted before military action began). In 1805, Napoleon had his primary base on the Rhine, in addition to several secondary ones (the first on the Lech, the second on the Inn, and the third in Vienna). While these bases were approximately eight days from one another, their creation (a very prudent action on the part of Napoleon) did not require any— even a minor—stopover of the French Army (on 20 October the French Army left from Ulm for the Inn river, and by 13 November the French already occupied Vienna, thus marching—with battles along the way—a distance of more than 500 kilometres in twenty-four days).

An attempt *to move from the analysis of strategy towards its synthesis* was made by the following writers:

1) The writings of *Joseph von Xylander*,[22] which, in fact, represent a mechanical compilation of the works of Bülow and Archduke Charles. He produced a list of *hard*, definitive rules combined with a speculative fantasy about the predominant significance of rivers and mountains over other territorial obstacles. According to his theory, control over the upper reaches of a river is especially advantageous. This claim raises the question: what were the benefits to the Allies in 1813 of controlling the river Elbe? On the other hand, it was very beneficial to Napoleon, who did not control its upper reaches.

---

[22] Joseph Carl August Anton Aloys Ritter und Edler von Xylander (1794–1854) was a Bavarian major general, authorised representative for the German Federal Military Commission and member of the Frankfurt National Assembly. He was also a military writer, historian and member of the Royal Swedish Academy of War Sciences.

2) *Antoine-Henri Jomini*, whose later works {His first work was titled *Tableau analytique des principales combinaisons de la guerre*, and his last *Précis de l'art de la guerre*.}[23] discussed the issue of military activities in relation to their moral, political and tactical elements. Moreover, he also assumed that a division of forces is possible, though only so far as the situation dictates, and only as long as the ability to reunite them at a decisive point is maintained. He is blamed for prioritising internal lines for no real reason, as generally he was not given to formulating definitive rules. Every topic discussed by him does not end with a hard and fast conclusion, but with a recommendation to study military history critically ('l'histoire, mais l'histoire militaire bien raisonnée'). Thus, the works of General Jomini, together with the works of Lloyd, deserve to take the place of honour in the strategic literature.

In addition to the writings of Lloyd and Jomini, the body of strategic literature can be strengthened by:

3) The work of *Carl von Clausewitz* ('Vom Kriege') is a good example, though the original author's plan was not accomplished and his language is too ambiguous. Like the works of Lloyd and Jomini, Clausewitz's work represents a synthetising treatise on strategy that decisively rejects any definitive systems or rules for military action. According to Clausewitz, the main purpose of theory is to offer a better understanding of military affairs.

4) The work of *Karl Wilhelm von Willisen*[24] {'Theorie des grossen Krieges'}[25] is another good example, though he neglected the moral element and was fixated on his desire to create a strict theory with a system of rules.

---

[23] Leer refers to Antoine-Henri Jomini, *Tableau analytique des principales combinaisons de la guerre, et de leurs rapports avec la politique des états, pour servir d'introduction au traité des grandes opérations militaires* (St Petersburg: Chez Bellizard, 1830); *Précis de l'art de la guerre, ou Nouveau tableau analytique des principales combinaisons de la stratégie, de la grande tactique et de la politique militaire* (Brussels: Meline, Cans et Compagnie, 1838)

[24] Karl Wilhelm Freiherr von Willisen (1790–1879) was a Prussian general.

[25] Leer refers to Karl Wilhelm von Willisen, *Die Theorie des großen Krieges* (Berlin: Duncker and Humblot, 1840).

There are also some authors who do not believe in a positive theory of strategy. These include:

5) *Karl von Decker*[26] did not try to produce any rules. By weighing up all possible pros and cons of military actions (i.e., analysing their characteristics), he simply tried to solve the main purpose of theory—to explain the essence of military activity.
6) The same can be said about the work of *Reinhold Wagner*, excepting his completely unnecessary part entitled *clean strategy*.[27]
7) *Napoleon* did not write any complete dogmatic treatise on strategy. His thoughts on the ways to wage wars are briefly discussed in his memoirs, and were collected in a separate book entitled 'Maximes de guerre: Pensées de Napoléon'. His whole theory is based on his idea that war should be waged according to the available means and given situation.
8) Another example was *Georg Heinrich Berenhorst*, who took an even more extreme position by attributing everything in war to chance, especially after the appearance of firearms on the battlefield.[28]

A similar denial of positive theory can also be found in other literature. Unlike the works already mentioned, this literature focuses on the study of *the most significant expressions of strategy*. These include: (1) *Baron Nikolai von Medem*;[29] (2) *General Modest Bogdanovich*;[30] and (3) *Friedrich Wilhelm Rüstow*.[31] The last wrote a brief description of the most significant contemporary wars.

---

[26] Karl von Decker (1784–1844) was a Prussian major general and a military and fictional writer.
[27] Leer refers to Reinhold Wagner, *Glavnyye cherty chistoy strategii* [The main characteristics of clean strategy], translated from the German by Fedor Tol' (St Petersburg, 1816).
[28] Leer refers to Georg Heinrich von Berenhorst, *Betrachtungen über einige Unrichtigkeiten in den Betrachtungen über die Kriegskunst, über ihre Fortschritte, ihre Widersprüche und ihre Zuderlässigteit* (Berlin: F. Nikolai, 1802).
[29] Baron Nikolai Medem (1798–1870) was a Russian general of German origin, who was a professor at the Imperial Nicholas Military Academy.
[30] Modest Bogdanovich (1805–82) was a Russian lieutenant general and military historian.
[31] Friedrich Wilhelm Rüstow (1821–78) was a Prussian-born Swiss soldier and military writer.

This is the history of the literature on strategy. However, what are the consequences of this colossal body of work?

Strategy, as defined by Lloyd, who planted it in *positive-scientific soil*, was predicted to have great success. The reality, however, turned out to be different. After Lloyd's initial impulse, many thinkers rushed in to develop strategy, without fully understanding or perhaps consciously neglecting his study. These scholars have produced an infinite number of one-sided (and, therefore, false) treatises. Focusing only on one element (be it administrative, tactical or terrain), these works elevated only one side of military affairs into the main principle, forcibly extending its influence to all military affairs. These manipulations have led to formalistic compilations (that unconsciously blend everything together), speculative fantasies, metaphysical or dialectical treatises, and, finally, a whole spectrum of denialists who reject the positive theory of strategy by tearing everything apart and suggesting nothing or very little instead. Generally speaking, while these secondary issues occupied the centre of the discourse, the true nature of strategy was pushed to the edges. In this way, many scholars, instead of dealing with the main problem, focused on various problems of definition, attempting to answer questions like: *Do military affairs have their own science, or only theory? What is strategy and what is tactics? How can they be differentiated?* {These questions, however, are not very substantial. In reality, strategy and tactics are two constantly fused sides in every military affair. It is impossible to draw a distinguishing line separating them. The main goal of theory should not be an attempt to separate them, but to show how they are infinitely interwoven in each military affair. If, when conducting a critical analysis of a military affair, we focus on its strategic side—this is *strategy*. If we focus on its tactical side—this is *tactics*. Let's see how the most significant writers define strategy and tactics. *Jomini* attributes to strategy all aspects of the theatre of war. Tactics, according to him, is the art of waging a battle. According to *Clausewitz*, tactics describe how to conduct battles, and strategy describes how a series of battles are combined to achieve the aim of a war. *Bülow* was probably the one who cared most about this question, and therefore solved it best. He described tactics as a science of military action against a *physical enemy*, and strategy as a science of military action against an *enemy's goals*. In other words, tactics is the science of all actions conducted within the enemy's view, and strategy is a science of all military action conducted outside the immediate view of the enemy. Finally, there is a third and the most curious definition given by Bülow: tactics is where people fight; where they do not fight, it is strategy. Bülow would not be far from the truth if he said: tactics is where people fight (*execute*), and strategy is where people combine (*contemplate*).}

While attempts to answer these questions have filled many books, the answers have not become clearer. The opposite is true, as each new author proposes new definitions and interpretations. When writing on *strategic combinations*, most authors have not bothered to grasp their soul, their inner side. Instead, they grasped their skeletons, focusing on only a few of the infinitely various forms that these combinations can take in each given situation. This has led to treatises on operational lines (parallel, concentric, eccentric) or on different shapes of the base (convex, concave, etc.). In this way, strategy has been filled with vague definitions, terms and arbitrary classifications (for example, fortifications of the first, second and third rank). Strategy has turned out to be about lines and angles. But the real aim of strategy is *to comprehend the true spirit of exemplary military operations*, to draw inspiration from their examples and develop the creativity that is so needed in any art, and particularly in military art (as it is more difficult and complex than other arts). It does not seem that when choosing his operational lines in 1800, 1805, 1809 and 1796, Napoleon was thinking about Bülow's angles, which were considered by Bülow as the highest expression of strategy, and which were mocked by Jomini as strategic pig's heads, in reference to the famous attack formation of the Greeks. These blunders did not stop there, and the evil continued to prevail. This method of analysis drove many weak and not so bright minds to suggest various unqualified conclusions and systems (such as: *Franz von Weyrother*:[32] a strategic diversion is *always* better than a tactical one; *Bülow*: the importance of flank positions; *Karl Ludwig von Phull*:[33] a defensive war has

---

[32] Franz von Weyrother (1755–1806) was an Austrian staff officer and general who fought during the French Revolutionary Wars and the Napoleonic Wars.

[33] Karl Ludwig August Friedrich von Phull (1757–1826) was a German general in the service of the kingdom of Prussia and the Russian Empire.

always to be conducted by two armies, where one army holds the enemy in the front, and another acts on his flank).

VI

On different forms to discuss the theory of military affairs. The abuse of the dogmatic form. The advantages of the historical method. The need to unite conceptualisation and history in discussing any science, i.e., the critical-historical method of discussion. The purpose of critical military history. The selection of events, their discussion and grouping: the three steps of any critical-historical work. Theoretical rules. Critical-historical analysis should be conducted in two simultaneous ways: (1) a detailed critical analysis of one military campaign; (2) an analysis of strategy and tactics based on examples.

The brief examination of the literature on strategy in the previous paragraph shows how weak the achievements of those who wrote on strategy have been: very ambiguous definitions, and arbitrary classifications, lines of argument and examination. Moreover, the worst thing is that many authors suggest absolute rules and systems, arguing for only one possible solution, while in fact there are an infinite number of possible solutions based on the circumstances of infinitely changing situations.

The main cause of this detrimental outcome probably lies with the *abuse* of the conceptual-dogmatic form of writing, over-theorisation, and a desire to submit the practical side of affairs to positive theory. {Is it possible to produce rules on how to write Homer's 'Iliad' or one of Corneille's tragedies?—Napoleon used to ask—'Obviously it is a matter of inspiration. The same can be said about the products of our art, military combinations (operations).' According to Lloyd, in military 'art, as in poetry and eloquence, there are many who can trace the rules by which a poem or an oration should be composed, and even compose, according to the exact rules; but for want of that enthusiastic and divine fire, their productions are languid and insipid: so in our profession, many are to be found who know every precept of it by heart; but, alas! when called upon to apply them, are immediately at a stand. They then can recall their rules, and want to make everything, the rivers, woods, ravines, mountains, etc., subservient to them; whereas their precepts should, on the contrary, be subject to these, who are the only rules, the only guide we ought to follow; whatever manoeuvre is not formed in these, is absurd and ridiculous.'[34]}

Obviously, the conceptual form of writing should be placed above others. However, everyone knows *that the best kind of conceptualisation is your own one, i.e., concepts that consist of your own observations and conclusions.* For an outside reader, even the best conceptual work, written by an excellent thinker, is nothing more than a collection of general statements. This is because it offers only the final products of the writer's thoughts, without providing the facts and observations that served to create these products, and through which they can be verified. Without these facts and observations, the only way to read these works is to trust the writer. Therefore, while accepting that the conceptual way of writing is the most accomplished one, we should keep in mind that it is not appropriate for an *initial* introduction to military affairs. In other words, we should approach conceptual work only *after* a comprehensive analysis has been made of everything related to military affairs, which brings together various strands of knowledge, obtained in various ways, in one systematic and harmonious whole. {Military theory can teach us only the knowledge required to understand everything related to military affairs. It does a much poorer job in teaching how this knowledge should be applied in practice. Military theory can be delivered in many different forms: a simple conversation; in the form of a poem ('Iliad'), a novel ('Cyropaedia') or a story; or in the form of dogmatic-conceptual works, in which a collection of rules represents the lowest level of these works, and a conceptual work represents the highest. [...] The main goal of various courses and books on tactics and strategy is the development of an understanding regarding the importance of initiative; the vital nature of surprise; the correct assessment of the environment; the ability to exploit the moment and use forces accordingly during the most decisive stages of the battle; and the important ability to comprehend moral-spiritual conditions.

A *poem* is probably the least successful form for achieving this goal, because it is too engaged with itself, appealing equally to the imagination and the mind. A common reader would find it difficult to uncover

---

[34] This quote comes from Lloyd, *Reflections on the General Principles of War*, pp. vii–viii.

the truth hidden by the poet between different imaginary characters and events. [...] However, there are some works of genius ('Iliad' or 'Faust') that can be excluded from this rule. After all, what is 'Faust' if not a conceptual treatise written in poetic form?

A *novel* gets the reader closer to his goal, as its author tries to give more concrete form to his ideas, forcing his characters to express and discuss them. However, it is also imperfect, as it tries to attract the reader by its superficial imaginary side, thus making it difficult to uncover the true meaning of the affairs presented.

A *story* can get the reader much closer to his desired goal, especially if its author does not incline towards imagination or any other attempt to improve the story at the expense of the actual events (*factual history*). A story would be especially useful if it described the events in much detail, allowing the reader to fully assess them, and if it presented the events in as lively a way as novels usually do (i.e., as a lively story and not as a dry skeleton). Therefore, the best story includes both lively description of the events in detail and the proper assessment of the events that offer instructive value to the reader (*critical history*). [...]

A *dogmatic-conceptual work*, unlike all other forms described above, resorts to neither imaginary nor real events or characters. It extracts the essence of the affairs and analyses them from multiple angles. It presents the embodiment of this essence, without concerning itself with its outer forms.

On the one hand, it seems obvious that the conceptual way of writing is the most accomplished one. On the other hand, as discussed previously, it is also obvious that it is not appropriate for an *initial* introduction to military affairs and it should be approached only *after* a comprehensive analysis has been made of everything related to military affairs. Otherwise, it will lead to an infinite variety of reasoning and all sorts of imaginary theories. The whole history of military literature supports the argument that conceptual writing is the most accomplished: the first military writer was a poet (Homer's 'Iliad'), who was followed by a novelist (Xenophon's 'Cyropaedia'), after whom came a historian (Polybius), and only later do we see the appearance of deductive-conceptual military authors. The same order can be also traced among the military writers of the Middle Ages or the Renaissance. That the conceptual way is the most accomplished is supported by the way each of us has developed in his life. The first notions about military affairs we get from conversations. And the first thing that usually strikes us most is the external and poetic side of military affairs. When reading about some famous battle, we have great pleasure returning time after time to its most exciting episodes. Our imagination is seized and our momentary obsession turns into a more solid concern. We start seeking a deeper acquaintance with the topic and find it in history, in the experiences of the past. We start to think about events, criticising them and making our own conclusions. Finally, driven by the desire to generalise these conclusions, to create one systematic and harmonious whole out of them, we try to extract their main fundamental roots (i.e., conceptualisation). This is the natural plan for studying military affairs, either by individuals or by the whole of humanity. It should start with the enjoyable poetic side, followed by the more serious historical one, and finish with the most accomplished conceptual side. For those who follow this natural way of self-development and reach the conceptual level, conceptualisation represents the crown of honour. This way, however, is not an easy one. The way of detailed microscopic analysis is boring and tedious. [...]

This is the main reason why the majority of writers rush directly into conceptual writing, bypassing the analytical aspects of military affairs. Conceptualisation not based on comprehensive analysis leads to an infinite variety of thoughts and insights. While some of these errant writers acknowledge their mistakes and, after a short adventure with conceptualisation, return to the true path of analysis, the vast majority persist in building the apex of their construction without solid foundations. For them, conceptualisation is not the crowning glory of military affairs, but a vacuous discussion of them.}

Strictly speaking, it is obviously impossible to prefer one form of writing entirely at the expense of another. The form itself is less important. What is important is that the essence of things should be communicated to the reader. Therefore, in an attempt to achieve this goal, it is vital to allow writers to freely choose any form of writing, according to their preferences, skills and ability to express their thoughts in an *accessible* way. Keeping this point of *accessibility* in mind, one must agree with Lloyd, preferring the historical form of writing and concluding that the most accomplished method is the one that combines both *historical* and *critical* forms of writing. {Comparing the benefits of deductive and historical ways of writing, Lloyd prefers the latter. According to him, 'the didactical kind communicates its precepts purely and simply, without any application: so that they make but a weak and transitory impression on the mind, which time soon erases. From whence it is become a proverb, that the road to science, by means of rules and precepts, is long and tedious; that by example, short and agreeable [*longum iter est per praecepta, breve et efficax per exempla*].'[35] 'A reader, no doubt, is much more attentive to real, than to imaginary transactions. He believes that it is possible, at least, to imitate what others have executed. There is besides, in every elevated mind, an emulation which encourages and animates us to tread the footsteps of those great men, whose actions and characters are justly the object of our love and veneration. For which reason, history has been ever recommended as the best, easiest, and most effectual method to instruct mankind.'[36]}

Truly speaking, if we wrote works of synthesis (treatises about a combination of elements) on strategy and tactics only in a dogmatic-conceptual way, we would be discussing exclusively ideas that are probably incomprehensible to many people, who are not practised in focusing their whole intellect on one specific topic (after all, laziness is one of the defining aspects of human nature). However, if we could supplement dogmatic-conceptual writing with an analysis of events and examples to demonstrate how our ideas were implemented in practice, then it would be an entirely different story. In other words, if in our studies of strategy and tactics we could combine both conceptual and historical ways of writing, we would make our ideas accessible to both those who are practised in concentrating their thoughts on one subject and those who are not. [...]

The benefits of critical-historical writing, however, are not limited to the fact that the expression of ideas makes them more accessible. Additional advantages of critical-historical analysis over conceptual include:

---

[35] This quote comes from Lloyd, *Reflections on the General Principles of War*, pp. i–ii.
[36] Ibid.

STRATEGIYA

1) Conceptual theoretical writings introduce ideas only and do not discuss practice. Thus, if all the foundations are not properly examined, these writings can easily turn into groundless empty words. While conceptual-theoretical work limits itself to *what should be done*, critical-historical analysis constantly moves between theory and practice, thus discussing *how it should be done* in a specific case.
2) Critical history objectively describes the influence of the situation on actions (as 'acting according to the situation' is the main rule of war). Thus, this analysis involves an understanding that the same actions, either offensive or defensive, must take infinitely different forms, depending on the situation, which by itself changes constantly. The ultimate conclusion of critical-historical analysis about rules is that 'everything is lawful in its time and place', and, therefore, it has a quite refreshing and sobering influence on the mind of the reader. {In *the 1805 Battle of Austerlitz*, an attack on two fronts was used incongruously, in defiance of the situation, thus ruining the whole affair. On the second day of *the 1813 Battle of Dresden*, as well as in *the 1866 Battle of Königgrätz*, the same method of attack was used, according to the situation, thus leading to the victory of the attacker. In *the 1859 Battle of Magenta* and in *the 1757 Battle of Prague*, it seems that the actions of the attacker were outrageously bold and against all possible theoretical-conceptual rules. However, taking into consideration the moral characteristics of the defenders, it turned out that the decisions of the attackers were right. These are good examples of *the meaning of rules* and *the influence of the situation*.} Theory leads towards a solid conclusion; practice, however, i.e., *military critical history, breaks it into a variety of infinitely different forms.*
3) Theory discusses a problem from the narrow perspective of one possible condition. Therefore, as discussed above, it can easily fall into a *dogmatic* preference for one solution out of the infinite variety of all possible solutions to this problem. In real life, however, everything happens under the influence of an aggregation of different conditions that shape the infinitely changing situation. *Focusing on this aggregation of different conditions and on the situation as a whole, critical-historical analysis rescues the reader from dogmatism.*

If, in an attempt to escape dogmatism, conceptual theory tried to do the same as critical-history, it would ultimately fail. Such an attempt requires an extraordinarily ardent imagination to create *a priori* assumptions regarding all possible forms of the situation. {Sometimes, even an unimaginably minor event can have a decisive impact. [...] Theory is incapable of taking such minor events or accidents into account. Critical history, however, can emphasise their importance, teaching us to respect accidents and be ready to face them calmly.} Even if theory could take up this task, everything discussed by it would be hypothetical and not factual, and any conclusions derived from hypotheses are not as reliable and trustworthy as the conclusions derived from facts. {Strategy created in this way would ultimately be a *hypothetical strategy*, i.e., some sort of *metaphysical strategy*. Critical-historical analysis, however, deals with facts and, therefore, creates *practical, positive strategy*.}

To summarise, military critical history embodies ideas, thus assisting their understanding, it helps to avoid unqualified assumptions and systems, rescues the reader from dogmatic conclusions and teaches him to respect the role of the situation, the true potentate of war. [...]

While understanding the importance of military critical history for the study of strategy and tactics (especially works that deal with a combination of elements), it is still important to ask: *how should it be discussed and studied?*

The main task of military critical history is to analyse an event in a way that will allow *the full comprehension of the ideas that stood behind it*, thus shedding light on the ideas that shaped it and re-creating the inner aspect of the facts related to it. The main difficulty is not in understanding *how* each event occurred, but in the full comprehension of *why* it occurred in one way and not another, in *the re-creation of the intellectual work* conducted by those who made decisions during this event. This re-creation would be the most helpful reading matter for anyone who is preparing himself for similar types of activities. This way of studying military history is the most suitable one, as it turns the process from an affair of *memory*, which it has mostly been so far, to an affair of the intellect. Events can be easily remembered if we comprehend the ideas that stand behind them.

While *selecting events* for critical analysis, it is important to take into consideration three main factors. The first one is *their educative potential*, either as positive examples of how one should act in a similar situation, or as negative examples of how one should not act under similar

STRATEGIYA

circumstances. From this perspective, the actions of such a brilliant general as Napoleon and the decisions of such a mediocre general as Karl Philipp, Prince of Schwarzenberg are equally instructive and, to a certain degree, equally exemplary.

Secondly, it is preferable to choose events not from the distant past, but from *as recent times as possible*. Events which are close to modern times introduce to the reader the contemporary state of military art and contemporary combinations of various elements. This is especially important, as the first and foremost duty of any officer is to be closely familiar with the contemporary state of military affairs. While a study of similar questions from the distant past is quite important, it is an exercise beyond what is necessary and should be conducted only after everything of primary importance is finished. With this in mind, there is no need to be too strict regarding the length of time that has passed since the event. After all, the time that has passed is not as important as the substance of the event. What we need is *paragons*. While these paragons are scattered throughout history, most frequently they can be found in the time of the great masters of art. The distinctive characteristic of paragons, regardless of the nature of their art, is that *they never get old*, but serve as eternal sources of inspiration for future generations. {Are the preparations of the theatre of war conducted by Gustavus Adolphus in 1630 and in the first part of 1631 less exemplary and instructive than the similar actions conducted by Napoleon in 1809? We dare to hope that they are not. [...]} In fact, neither strategy nor tactics has changed throughout time. In all times their main goal was *to find the most advantageous way to combine elements* (of the theatre of war—for strategy; and of the battlefield—for tactics), *while achieving a defined goal in the shortest possible time and with the least sacrifices*. Both Genghis Khan and Napoleon tried to achieve this goal; they only did it differently. While strategy and tactics are unchangeable, what constantly changes is their means, the situation in which they are used and, obviously, the forms of the military combinations deployed.

An additional important factor that has to be taken into consideration, while choosing events for critical analysis, is *the requisite variety in their situations*. After critically analysing one event, it is advisable to move directly to another, which is very similar in character but differs significantly in situation. The larger the difference the better, because it will help to establish the primary importance of the situation and avoid unqualified dogmatic conclusions. The best choice is when the difference between two situations is so wide that the means responsible for success in the first event was used in the second incongruously with the situation, leading to ultimate failure. The conclusions derived from the study of the first event would be negated by the analysis of the second, and thus they cannot be used as unqualified lessons for other similar cases.

Once events are chosen, the next question is: *what should be emphasised in their critical analysis?* As discussed above, the main goal is to comprehend the idea that lies at the heart of the event. In doing so, it is necessary to re-create, *as fully as possible*, the situation in which this event occurred. In other words, one has to take into consideration the specific combination of *all* circumstances and *all* causes that influenced the event to occur in the way it did. In short, one has to re-create everything that was thought and felt by those who made the decisions in this event. Only this way of studying military history can offer real value. Only conclusions derived from this way of analysis are like conclusions obtained in practice. Finally, only this way of analysing events gives one the right to make conclusions at all. Otherwise, if the analysis is based only on a partial picture, completely overlooking some circumstances and leading to an incomplete re-creation of the situation, then all conclusions derived from it would be very one-sided and, therefore, false.

Once events are chosen and analysed, the next question is: *how should they be collected into one treatise?* A collection of critical analyses of different events without any *logical connection* between them seems inconceivable.

The easiest way to achieve this connection is by critically analysing different events that are parts of *one* of the most remarkable campaigns. While this way has its own advantages, it also has one significant disadvantage. After all, in one campaign all solutions of strategic (and, obviously, tactical) problems will take a single form, thus leading the reader to consider it the best one. Therefore, in this way, one of the major goals would not be achieved—an understanding *of the infinite variety of possible solutions*, which depend on the infinite variety of situations. {As, according to Napoleon, 'all questions of higher tactics (strategy) seem like indeterminate physical-mathematical problems that allow multiple solutions'.}

To achieve this goal, it is necessary to analyse as many similar decisions as possible. It is important to keep in mind that these decisions should follow the same order as they were taken in war (such as drawing up a plan, choosing the nature of the action, creating operational lines, concentrating supplies and forces, combining manoeuvres, fighting, etc.). Moreover, as mentioned above, it is important to focus on similar decisions taken under *many different conditions*, thus obtaining as many different solutions to one problem as possible.

At the root of every decision, as discussed in paragraph III, lies a collection of several factors that this decision has to meet (theoretical rule, law, general formula, unchanging ideas, Napoleon's axes for plotting curves). Therefore, a critical analysis of any strategic issue should include three major steps. Firstly, it must *define* the rule that lies at the heart of this issue and prove its correctness by means of analysis. Secondly, it must *study* it, as one studies general mathematical formulae. {For example, in mathematics, an equation of second-order curves, under different assumptions, can become a circle, ellipse, hyperbola, parabola, etc. Similarly, in our case, the theoretical rule can be implemented differently under different circumstance. In other words, a variety of forms is possible, when there is no absolutely best form for all cases, though there is one absolutely best for each known case.} It has to show what form a theoretical rule can take in one or other situation. {Let's take, for example, the theoretical rule regarding defence positions, i.e., the aggregation of conditions that is deemed by science as the best defence position. It is obvious that in practice there is no such position that would fulfil the aggregation of all possible conditions and circumstances. Therefore, this rule should be a theoretical ideal of defence positions, which, similar to the example discussed above of mathematical formulae, does not define a single best position, but formulates the idea of all possible ones.} There is no doubt that the number of possible forms is infinite, because the number of variations of conditions is infinite as well. [...] However, regardless of this infinite variety of different answers to the same question, each one of which is shaped by a unique aggregation of conditions, the theoretical foundations (rules) that lie at the heart of this question are unchanging. While two different orders for battle might be as different in form as the difference in clothes of a giant and a dwarf, they are very similar at their core—since, regardless of size, both giant and dwarf wear clothes.

This is why after deconstructing theory into separate cases, one has to move to the third step—*the identification of characteristics that are shared by all the cases or, at least, by the majority of them*. In this way, we will come back to the theoretical foundations that served as the starting point of research. We will close the circle of research that starts from the theoretical idea, moves through its practical (infinitely different) forms, and returns to the theoretical foundations, ideas that generalise the infinite number of solutions to the question of our research.

This way of inquiry *must* be used in the analysis of all questions of strategy (and tactics) related to the combination of elements, because only this way can offer the correct view of military affairs and rescue us from the disastrous focus on a single possible solution. [...] In sum, only this way will take strategy, which, like other sciences, has already passed through its *metaphysical* and *scholastic* periods and currently exists in its dialectical phase, and turn it into *a positive science*—a science that is based on unchangeable laws, which are derived not from the depths of the writer's soul, but from the critical analysis of history and practice. {As this treatise is the first-ever (and therefore quite weak) attempt to implement this way and take strategy from its dialectical phase towards positive science, we will permit ourselves to call it *positive strategy*. This name corresponds to the main task of this treatise in comprehending the primary characteristics of strategy not by hypothesising about it or producing metaphysical ideas, but by way of critical analysis. [...]}

Most of the authors on strategy have adopted the dogmatic style of writing, thus ultimately falling into a very dangerous trap. They have given preference to a single specific solution, arguing that it is universally true and can be applied in any situation (out of the infinite variety of possibilities). For example, discussing the question of the base of operation, *Dietrich Heinrich Freiherr von Bülow* argued that the base should be long and concave. *Archduke Charles, Duke of Teschen*, however, discussing the same question, claimed that it should be long and parallel to the enemy's base. While *Joseph von Xylander* stated that the base of operations should be concave and slightly offset from the enemy's base, *Frédéric François Guillaume de Vaudoncourt*[37] argued that it should be perpendicular, and *Antoine-Henri Jomini* advocated that it should be double—parallel and perpendicular at the same time. As the reader can see, every possible solution has been proposed by these writers. And while each of them seems to offer a good solution to a specific situation, the main problem is that each writer advocates his solution as a *universal prescription*, which it is definitely not.

Those who argue for universal prescriptions miss the fact that *theory*, as Napoleon used to say, offers only *the axes that allow the drawing of curves* (theoretical rules). {As discussed above, theoretical rules are, in fact, these *axes*. Therefore, theory offers *axes* that help one to decide about the order for battle, to choose positions, operational lines, etc. However, it cannot offer the best order, position or line for every given situation (i.e., the *curves*, according to Napoleon)—this is the domain of practice. In other words, *the axes are given by science, but drawing the curves is a matter of art.*} Moreover, we will add that theory cannot explain how to draw the curves; this can only be understood from practice and history. In other words, *practice draws one or other curve according to the specific conditions of the situation, utilising theoretical axes* (theoretical rules).

By giving ultimate preference to a single specific solution, each of these authors gives to his theoretical curve (which is right for one specific situation only) a status that no theory should claim for itself. If a theoretical treatise wants to stay true to itself, then it should avoid giving life to these pitiful universal prescriptions, which not only fail to comprehend the essence of strategy, but also confuse and bedevil people with limited capabilities and imagination. Because, while some of these people, influenced by this deceptive theory, spend hours trying to find a place for the universal base of operations (which simply does not exist in practice), others take up the idea of a manoeuvre, or a flanking position, or a secondary battlefield order, and repeatedly deploy it, in defiance of the conditions of the situation, thus ultimately

---

[37] Frédéric-François Guillaume de Vaudoncourt (1772–1845) was a French general, writer and military historian.

causing hundreds of thousands of innocent casualties. {These people, who studied a lot without really grasping the essence, were famously called by our great Suvorov *poor academics*. He writes about them in his famous plan against the French Republic: 'Off with methodology, leave manoeuvres, marches and counter-marches to the poor academics' (obviously as an idea of universal prescriptions, not as actions in accordance with the requirements of situation). [...]}

In our opinion, the only way to get rid of this significant disadvantage of current literature is to investigate the source of these false, one-sided conclusions. Apparently, finding this source is not so difficult, because it is rooted, as already mentioned above, in *over-theorisation* and dogmatisation of the practical sides of strategy and tactics (in the desire to *draw curves*, i.e., to teach something that no science can teach). In an attempt to get rid of this disadvantage, it is important to *push theory away*, strictly defining what it can describe and what not. In the process of combining elements, theory should be given the right to define *axes* (theoretical rules) only. Then, it should be complemented by *as wide an analysis of practice as possible*, i.e., by the military critical history that analyses as large a number of different *curves* as possible in as many different situations as possible. This is the essence of the *critical-historical form* of analysis—the idea of a merger between *historical and conceptual ways* of research into strategic affairs. [...]

VII

The one-sidedness of the majority of historians. The necessity of a new school that will emphasise the comprehensive description and analysis of events.

Previous studies have already emphasised the importance of military history for the study of military affairs. For a military person, military history is like a *laboratory* for a chemist or an *observatory* for an astronomer. Military history, combined with experience, as Napoleon used to say, is the *only* tool to discover the secrets of the higher reaches of military art.

It is important to note that, thanks to the serious mistakes made by the majority of authors in the field, this sole tool is far from where it should be. While there are many mistakes made by these authors, their *one-sidedness* and *imprecision* are, undoubtedly, the most serious ones.

STRATEGIYA

This reproach, which can rightfully be extended to many contemporary historians {with the exception of a few outstanding military-historical works that appeared lately in Russia, as well as abroad}, was made long ago by Lloyd. According to him, 'Those historians, both ancient and modern, who have given us an account of different wars, though in many respects extremely valuable, are not as accurate as they might, and ought to be.'[38]

Following this general accusation, Lloyd continues: 'They do not describe with sufficient precision and exactness, the countries wherein the wars were carried on, nor the particular spots upon which some great transactions happened; the number, species, and quality of the troops which composed the respective armies are generally omitted, as well as the plan of operation; and the operations themselves, excepting those which appeared extraordinary. They do not explain minutely, as they ought, why, how and where every operation was transacted. They only, in general terms, give the history of a campaign, without explaining sufficiently the motives by which the generals were actuated, how the various operations of it were conducted; and lastly, what was the nature of the ground where they happened.'[39] While Lloyd concludes by saying, 'the knowledge, however, of these points is so necessary, that it is impossible to form an exact opinion of the propriety or impropriety of any military transaction without it,'[40] we would add that the lack of this knowledge also nullifies any possible value that could be achieved from studying military history.

It is fairly obvious that Lloyd was quite right in his identification of the main faults of the majority of military-historical works, such as imprecision, one-sidedness, dry presentation and general neglect of a serious kind. To demonstrate the correctness of Lloyd's reproaches, it seems important to discuss them further.

It seems not so difficult to recognise the *one-sidedness* of many military-historical works. The vast majority of them consist of descriptions of battles and manoeuvres. But, how many of them include a compre-

---

[38] This quote comes from Lloyd, *Reflections on the General Principles of War*, p. ii.
[39] Ibid., pp. ii–iii.
[40] Ibid.

hensive description of the theatre of war or various engineering or administrative measures conducted to prepare it? How many of them include detailed information about the organisation of the fighting armies or the moral characteristics of the commanding generals on both sides? How many include a detailed description of the relations between the military commanders and their respective governments, the relations between the states involved in the war and other states, or the way in which political conditions influenced military activities in general? However, without all these accounts it is simply impossible to fully comprehend the event, it is impossible *to comprehend the idea that stood behind it*—the essence of a proper study of military affairs. [...]

Military historians frequently fail to discuss the influence of political conditions on the course of military actions. This is not only an unforgivable mistake, but it can also lead to very incorrect conclusions. For example, in analysing the actions of Frederick the Great during his 1761 campaign from a very narrow military point of view and neglecting the political conditions, it is easy to conclude that his actions were wrong. From the narrow military perspective, he missed the opportunity to crush the army of Alexander Buturlin, which was vulnerable because it was divided by a river. However, once we re-create the whole situation according to which Frederick the Great made his decisions (i.e., taking into consideration as well the political conditions), we can clearly recognise that while his decision initially seems wrong, in fact it was an exemplary one. The political conditions shaped the situation in such a way that Frederick realised that crushing Buturlin would be like crushing himself.

If we analyse Napoleon's retreat from the 1813 Battle of Leipzig from the pure military (tactical) perspective, it cannot withstand criticism, as his actions violate even basic rules of tactics. However, from the political point of view, they are nothing but correct. From the military perspective, Napoleon should have withdrawn first the forces that were closest to Leipzig and ordered them to hold the city. Then, under cover of the forces in the city, he should have marched the rest of his forces through Leipzig to Lindenau, leaving a small detachment covering the retreat of the forces holding Leipzig. Napoleon, however, acted in the completely opposite way. He started the withdrawal of the forces closest to Leipzig (Victor, Augereau, the Guard), ordering them to march *directly* to Lützen. Then he ordered other forces close to Leipzig (Marmont, Dieu Soult, Lauriston) to similarly pass through Leipzig and retreat *directly* towards Lützen. The last to retreat from the battle were the forces of Poniatowski, MacDonald and Reynier. They were ordered to hold Leipzig for twenty-four hours. Since the forces that started the retreat were ordered to march *directly* to Lützen, it seems obvious that the forces left to hold Leipzig were simply sacrificed.

From the military perspective, Napoleon's orders seem to be quite wrong. However, an examination from the political angle reveals that they were very correct, though not necessarily fair. The critical situation in which Napoleon found himself required him to save the French forces at all costs. To achieve this goal, Napoleon did not hesitate to sacrifice foreign forces—the Polish corps (Poniatowski) and the corps of the Confederation of the Rhine (MacDonald), which he would lose anyway owing to his retreat westwards to the Rhine.

We hope that the examples discussed above are sufficient for the reader to realise how a *one-sided* description of military history can lead to wrong conclusions.

Generally speaking, military-historical research should find a new way of analysis which is not limited to the description of marches and battles. Instead, it should include as comprehensive an analysis of the events as possible. It should go as deep as necessary into the analysis of political and administrative factors. It even should focus on intrigues that usually occur in the headquarters of large armies, which, unfortunately, play an important and not necessarily honourable role in military actions. In general, military-historical research should start exposing the *inner* side of events, instead of superficially describing their *facade*. Otherwise, a comprehensive re-creation of the event is simply impossible—it would remain forever *a dead corpse*, and would never become *a live story*, the way military history (and history in general) should be written. Otherwise, it is entirely useless.

Some of the greatest contemporary authors have already realised the need for reform in historical research. For example, this is how this subject is discussed by Thomas Babington Macaulay, one of the writers of the new school, in his introduction to *The History of England*: 'I should very imperfectly execute the task which I have undertaken if I were merely to treat of battles and sieges, of the rise and fall of administrations, of intrigues in the palace, and of debates in the parliament. It will be my endeavour to relate the history of the people as well as the history of the government, to trace the progress of useful and ornamental arts, to describe the rise of religious sects and the changes of literary taste, to portray the manners of successive generations and not to pass by with neglect even the revolutions which have taken place in dress, furniture, repasts, and public amusements. *I shall cheerfully bear the reproach of having descended below the dignity of history, if I can succeed in placing before the English of the nineteenth century a true picture of the life of their ancestors.*'[41]

According to the old school, a focus on such *squabbles* is nothing more than an attempt to dethrone science and trample it into the dirt. However, this is exactly our goal, which our generation has understood perfectly. Science should be toppled from its throne, on which it climbed to no purpose (metaphysics, scholasticism, etc.), so that it will stop existing apart from real life.

This is the idea that should drive changes in military-historical research. The authors should not scruple to dig deeper into the inner side of events. After all, the only thing that they will find there is truth, though it will not necessarily wear the same finery as the event's outer appearance.

VIII

The creative and technical sides of military art in general, and strategy in particular. The organic connection and consistency of strategic operations, as well as their relative importance. Strategic terminology.

Military art, like any other art, consists of two parts. While the first part is based on *knowledge*, the second one is rooted in *aptitude*, i.e., creativeness. The first part is *elementary knowledge* (i.e., it gives basic rules and guidelines); the second is the sublime ability to *combine and create* (la partie sublime de l'art, la partie divine).

The first part, according to Lloyd, is the *material* side of art, i.e., the organisation and study of its elements and their characteristics, the

---

[41] This quote comes from Thomas Babington Macaulay, *The History of England from the Accession of James the Second*, vol. 1 (Leipzig: Bernhard Tauchnitz, 1849), p. 3.

so-called technique of art. The second, Lloyd claims, *consists of the ability to apply common guidelines* (rules) *to infinitely varied situations in a correct and firm way*. Neither history, nor the persistent study of it, nor experience (as rich as it can be) can teach this part. Understanding it is a matter of genius, and thus a matter of natural talent. [...]

As strategy is the art of waging war, it, like all military art, consists of two main parts: *elementary* and *creative*.

Of all the strategic questions in war, there are two important ones, whose solution requires the highest degree of creativeness. The first is *the choice of target for attack* (the enemy's most vulnerable side), i.e., *the object of action* or, as some authors call it, *the decisive point* or *the enemy's centre of gravity*. Strategy should also *formulate the desired aim* and *the line of operation*, as the object of the action emphasises only its final point. The second question, whose solution requires the highest degree of creativeness, concerns the *security of the chosen operational line* and the ability to protect it for the duration of the whole campaign.

Solving these two questions is, in fact, *the highest, creative part of strategy*. The remaining questions either prepare the ground for or supplement the solution of these two principal questions. Answers to these secondary questions lie with *strategic technique* (or, perhaps, *strategic mechanics*), which also requires a certain degree of creativeness (after all, what does not require creativeness?), but not to the same degree as the two principal questions. [...] Since the solution of the two principal questions involves nothing mechanical, it is based solely on creativeness and inspiration. [...]

While our view of the extreme importance of the principal questions of strategy might be formulated slightly differently, it generally falls in line with the views of *Lloyd*, *Jomini* and *Marmont*.

For example, *Lloyd* suggests one pay special attention to the choice of the operational line. According to him, the operational line 'of all those we have mentioned, is the most important. For on the good or bad choice of this line the final event of the war chiefly depends. If it is ill chosen, all your successes, however brilliant, will, in the end, *be found useless*.'[42]

---

[42] This quote comes from Lloyd, *Reflections on the General Principles of War*, p. 134.

'If the art of war', states *Jomini*, 'consists in bringing into action upon the decisive point of the theatre of operations the greatest possible force, the choice of the line of operations, being the primary means of attaining this end, may be regarded as the *fundamental* idea in a good plan of a campaign.'[43]

Marmont tasks strategy with a twofold objective. The first is 'to reunite all the troops, or the greatest number possible, upon the point of conflict, if the enemy has there but a portion of his; in other terms, to manage to secure a numerical superiority for the day of battle.'[44] This can only be achieved by a skilful *choice of operational line* and a similarly skilful population of this line by forces; in other words, *a skilful execution of a manoeuvre*, which, in fact, embodies the idea expressed by the operational line. {'Skilful marches', states Jomini, 'are but an application of the great principle of throwing the mass of the forces upon the decisive point'—*as the capacity to choose these points is the embodiment of art*. 'What was the passage of St Bernard but a line of operation directed against an extremity of the strategic front of the enemy, and thence upon his line of retreat?'[45]} The second objective of strategy, according to *Marmont*, is 'to cover and assure one's own communications, while all the time menacing those of the enemy'.[46] This issue is resolved by the *ability to safeguard your operational line for the whole duration of the campaign*.

The first and most critical decision that should be made before starting to draft *a campaign plan* is to *define its aim* and choose *the object of action* and the most suitable *operational line*.

It is important to note that the operational line should be understood as *an abstract line* (this is why the whole debate, whether it should be a line, zone or any other shape, is a complete nonsense). This idea of an

---

[43] This quote comes from Antoine-Henri Jomini, *Art of War*, translated by G.H. Mendell and W.P. Craighill (Philadelphia: J.B. Lippincott and Co., 1862), p. 114.
[44] This quote comes from Auguste de Marmont, *The Spirit of Military Institutions*, translated by Frank Schaller (Columbia: Evans and Cogswell, 1864), p. 38.
[45] This quote comes from Jomini, *Art of War*, p. 139.
[46] This quote comes from Marmont, *The Spirit of Military Institutions*, p. 38.

operational line, as noted above, is embodied in the *manoeuvre*, which for the most part will be no more than *a preparation for battle*. Only the battle itself should be understood as *the final decisive means*.

The *organisation of the army* and the preparation of *the operational base* for the concentration of *the required forces and supplies* should be carried out in the context of the defined aim and the characteristics of the theatre of war. {*Operational base* is a certain area of the country (usually near the border) used to concentrate all means required to meet the needs of the army and its reinforcement. Similarly, in the case of defeat, it should offer resources for the organisation of defence.} [...]

Strategic terminology, like professional terminology in any science, has its own, very important meaning. {Scientific terms and names should be treated with great respect. A term represents the comprehensive meaning of the phenomenon it describes. In other words, *if you name and call something correctly, it means that you understand it correctly*.} It is important to avoid abusing terminology, something that some people frequently do, especially those who incline to hide their intellectual vacuity behind scientific terms. Unfortunately, in the context of the current state of strategy, it is difficult to deny the fact that our strategic terminology is quite obsolete.

IX

Situation and its significance.

In concluding this part, it seems important to emphasise again one of the most important questions about the significance of situation in war (and in general life as well, because, after all, war is only one type of human social activity and, therefore, the same rules apply to the foundations of both our social life and war).

We have already mentioned several times the main and only rule given by Napoleon, that '*circumstances* (situation) *govern war*' (à la guerre ce sont les circonstances qui commandent). The true nature of this rule has been proven by many wars, as well as other activities in our social life. From our experience, we can see that the only actions that achieve their aims are those conducted in strict accordance with the situation. More than anyone else, great leaders remembered that rule and acted accordingly. This is why they were great. They had a naturally developed ability to assess the situation, regardless of the way it manifested itself, and skilfully combined all available resources to achieve their aim according to the specific characteristics of each given situation. While a great leader uses his natural talent of judgement and his creativity, lesser people come with rules, recipes and systems that suit only certain situations (and not all those possible, as they claim), thus hoping to comprehend the infinite variety of the phenomena of life with the help of lifeless, mechanical tools. It seems that these people fail to comprehend that the infinite variety of the phenomena of life requires a similarly infinite variety of intellectual creativeness. Lifeless systems and rules will be helpful only if, strangely, life becomes frozen in a situation that suits these rules and systems. Therefore, it is obvious that *there are no rules for action and there cannot be any rules for action* in the creative, practical parts of strategy and tactics, even though they are proposed all the time. Moreover, unfortunately, these rules frequently fall onto the very fruitful soil of fresh and young minds. Since any *first impression* usually has a very strong and lasting power, these rules grow very strong and deep roots, to a degree that someone who mastered them at a young age will follow them for the rest of his life. This education spoils the whole lives of these young people, guiding them in ways contrary to the nature of life. After all, *one who simply bows to the rules does not respect the significance of the situation*. Therefore, the logical opposite conclusion that should form the heart of education is: *one who respects the significance of a situation does not recognise the existence of rules*.

If we accept the fact that situation is the main determinant of war, then we should also accept that *actions can be considered exemplary only if they are conducted in strict accordance with the situation*. Therefore it is impossible to judge whether actions were 'sensible' or 'lacking in sense', 'right' or 'wrong', *without taking the situation into consideration*. [...]

This discussion about the significance of situation is lengthy and detailed for two main reasons. Firstly, because of the tremendous role that an understanding of situation plays in war. Secondly, because of the lack of respect shown to it by the majority of authors and practitioners alike, who have been raised on dogmatic treatises that exalt rules. [...]

2

THE RESPONSIBILITIES OF POLITICS IN ITS RELATIONS WITH STRATEGY[1]

Evgeny Ivanovich Martynov

In anticipation of those long-desired times when diplomacy will find a way to abolish armed clashes between peoples, it would be very useful to turn attention to something which has always been practically feasible—how to make wars more meaningful and less costly.

The resolution of this task lies, first and foremost, with diplomacy itself. War is only an instrument of politics, and therefore the latter not only determines the beginning of the former, but also has a very strong influence on the strategy employed.

This book attempts to clarify the responsibilities that politics must fulfil in its relations with strategy, in order to enable it to achieve the desired result in the shortest time and with the minimum of costs.

---

[1] This chapter is based on the translation of selected chapters from Evgeny Martynov, *Obyazannosti politiki po otnosheniyu k strategii* [The responsibilities of politics in its relations with strategy], (St Petersburg: Tipografiya Glavnogo Upravleniya Udelov, 1899).

STRATEGIYA

I. Politics must create the right guiding idea for its actions and choose allies and opponents according to it

Each nation, in any given period of time, has its own political idea. For one nation this idea is national unification, for another the creation of strong natural borders. For some nations it is about ruling the seas, for others it is about controlling profitable markets. These ideas are as various as the conditions of life in different states, as well as the needs and preferences of their peoples. People live and fight, achieving their desired aims or disappearing in the struggle for them. However, an aim once achieved will be always replaced by a new one. A state, to put it simply, cannot be a healthy organism without having a political idea. If we see a state that has no future political goals, nothing to desire and fight for, then we can be sure that this state has already fulfilled its role in history, and is in decline, living through a period of decay.

However, the political ideas of nations are not always defined clearly. There are times when one single idea unites a whole nation. However, too often there are times when the desire for personal peace and various cosmopolitan theories obscure a healthy national interest. Therefore, the first main aim of good politics is to fathom this interest, even in the maelstrom of many opposing currents, turning it into the guiding idea of all political activities. Firmly following a chosen direction, politics should eliminate any obstacles on its way, using all possible means, such as persuasion, negotiation, deception and, in very extreme cases, even the means of war. In this way, in the hands of good politics, war is not meaningless carnage, but the last weapon of choice for achieving a conscious aim. Each separate campaign brings the desired political idea one step closer, serving as an intermediate stage along the chosen path. A clearly understood guiding idea connects the wars of a particular period into one elegant whole. Several subsequent generations will labour to solve one clearly defined political question, and once its aim is finally achieved, human history records another logically comprehensible cycle of wars.

Only this type of politics can lead to great outcomes. The best example of this was the politics of Rome.

From the period of its mythical history and the two-centuries-long struggle between patricians and plebeians, ancient Rome emerged as a republic based on the equality of all classes. This happened in the middle of the 4th century BC, when the Roman territory was limited to a tiny region of Latium, whose borders are difficult to find on the modern map of Europe. Four and half centuries later, however, we can see the vast empire of Augustus, consisting of hundreds of millions of inhabitants and extending from the Atlantic to the Euphrates and from the Danube to the Sahara desert. How can we explain this extreme growth of Roman power?

On the one hand, it can be explained by the favourable central geographical position of Rome in the ancient world, the natural abilities of its citizens, the wisdom of its laws, and the strength of its military system. On the other hand, it does not seem that all these significant advantages of Rome could serve as the main reason for its success. After all, the Greeks were very close to the Romans, their geographical position was even better, they had skills and were gifted to the same degree as the Romans, if not more so. Greece, however, failed to create a strong, long-lasting state. The real reason for Rome's success lies with its strictly national, remarkably consistent and persistent politics, which were based not only on the above-mentioned advantages, but also on the extremely skilful use of its most decisive weapon—strategy.

The Roman rise to power lasted for three and a half centuries, from the moment when the Roman Republic resolved its internal affairs (in the middle of the 4th century BC) and entered the world-historical scene, until the rule of the emperor Augustus, when Rome reached its greatest power. While it might be seen as one long period, in fact it can be divided into five separate cycles of war.

The first cycle consists of wars for the unification of Italy. They started from approximately 350 BC with the wars against Rome's Samnite neighbours and ended in the middle of the 2nd century BC with the Roman conquest of Taranto and other states in southern Italy. These political actions of Rome lasted approximately a hundred years. The result was the complete conquest of Italy.

The second cycle consists of the Punic Wars—a series of fierce struggles between Rome and Carthage that started in 264 BC and ended in 146 BC. On several occasions Rome was on the brink of extinction, but after 120 years of effort, Rome finally succeeded in bringing under its rule all the lands on the western side of the Mediterranean Sea and some on its southern side.

STRATEGIYA

The third cycle consists of wars waged for the conquest of the Balkan Peninsula. They started in 195 BC and ended in 146 BC; hence this cycle coincided with the second part of the previous cycle. However, this simultaneous pursuit of different political aims does not constitute a political mistake. It was the Macedonian king Philip V who decided to intervene in the struggle by taking the side of Carthage against Rome. Following the end of the Second Punic War, when the power of Carthage was broken and its very existence lay in the hands of victorious Rome, the Romans decided to use the long pause between the Second and Third Punic Wars by punishing the Macedonian king and expanding their power to the Balkan Peninsula. Through a series of campaigns, Rome conquered Macedonia, Illyria and the Greek states. This series of wars lasted for almost fifty years, at the end of which Rome achieved full control over the Balkan Peninsula.

The fourth cycle consists of wars for control of the south-east coast of the Mediterranean Sea. The first step already took place in 190 BC, after the Second Punic War, when Antiochus III the Great was defeated at the Battle of Magnesia and was forced to give up significant territory in Asia Minor. On this occasion, this achievement satisfied the Romans, as they were busy with the war against Carthage and the conquest of the Balkan Peninsula. When these wars ended in 146 BC, new obstacles diverted the attention of Roman politics from the south-eastern territories. The uprising of conquered Spain, the revival of the internal struggle between patricians and plebeians, the Jugurthine War, the invasion of the Cimbri and the Teutons, a very difficult war with Rome's previous allies—all these distracted Roman politics from pursuing its aims on the south-east coast of the Mediterranean Sea. In the middle of these troubles, the only achievement of Rome in the region was the acquisition of Pergamon, which was bequeathed to Rome by Attalus III after his death in 133 BC. However, once the internal struggles were resolved, Rome's politics immediately returned to their original idea, answering the challenge brought by Mithridates VI of Pontus. In 84 BC, Sulla enforced peace upon Pontus and, twenty years later, Pompey the Great conquered the kingdoms of Pontus, Armenia and the Seleucids. Finally, in 30 BC, Octavius declared Egypt one of the Roman provinces. This is when the fourth cycle ends. From the campaign against Antiochus III until the conquest of Alexandria, this cycle lasted for approximately a hundred and sixty years.

While, during these four cycles, Rome's politics focused on the littoral of the Mediterranean Sea, there was another area that so far had preserved its independence. This was Gaul, and the Gallic Wars constitute *the fifth cycle* of Roman wars. This cycle included several campaigns conducted by Julius Caesar from 58 to 50 BC, and it was concluded with the complete conquest of Gaul.

This is how in slightly more than three centuries Rome conquered the whole known world. From a tiny Italian republic, it turned into a hegemonic state. The power that helped Rome in its achievement was *politics*, the main weapon of which was war. Rome's rapid and enormous success can only be explained by the skilful use of this weapon by its politics.

The first aim of Roman politics was the unification of Italy. Once Rome defined this political aim, it persistently and methodically marched towards it. All wars waged during this period are rooted in this aim, they are all connected by it into one elegant whole, similar to the way separate battles are conducted within a skilfully executed strategic operation. All subsequent periods also demonstrate the same characteristic of Roman politics. In its desire to rule the world, Roman politics marched in this direction by defining one aim after another and achieving them by a series of consecutive wars. On the one hand, the invasions of barbarian tribes, internal troubles, and the attacks of neighbouring countries interrupted the path of Roman politics. On the other hand, once these difficulties were resolved, it quickly returned to its old path. Rome's wars had always pursued only national interests. They neither aimlessly wasted national resources, nor were they conducted by accident. During each period, sometimes during several centuries, Roman wars were connected by a guiding idea, creating in time fully complete cycles.

Another very good example of perfectly complete cycles of war led by a firm political idea can be found in Russian history. This concerns the efforts over almost three hundred years of the Russian princes, who brought together the Russian lands.

Finally, the recent wars waged by the Germans have also very similar characteristics. Their main political idea was the unification of Germany under Prussia. The first step on this path was the Second Schleswig War in 1864, in which Schleswig and Holstein were seized from the kingdom of Denmark. The second step was the 1866 Austro-Prussian War, which led to the establishment of the North German Confederation, incorporating Hanover, Hesse-Cassel and Hesse-Nassau. Finally, the 1870–71 Franco-Prussian War allowed Germany to take back the old German provinces of Alsace and Lorraine. If the young German Empire succeeds in future in extending its rule to the German region of Austria, then its political aim will be fully accomplished, and all wars waged for this aim will be connected by one shared idea.

While learning from these positive examples is important, it is also important to learn from mistakes. There are two most common mistakes: either *politics has no guiding idea* or *this idea is wrong*. In the first case, even the most victorious wars will not lead to significant results, because politics pursues divergent aims and frequently destroys today what was achieved yesterday.

One of the best examples of this was the Russian politics of Elizabeth of Russia and Peter III. During the Seven Years' War, either following the advice of her chancellor Bestuzhev-Ryumin regarding the need 'to weaken the power of the fast-rising Prussian king', or driven by her personal disaffection with Frederick the Great, Elizaveta Petrovna, the Empress of Russia, decided to join the powerful coalition against Prussia. From 1757, Russian forces marched across Prussian territory, sometimes beating the Prussian forces and sometimes being beaten by them. After four long years of meaningless efforts (though they were sometimes very skilfully executed), Russian forces finally brought Prussia to the brink of defeat. Eastern Prussia even swore allegiance to Russia and had coins minted with the image of the Empress. At this exact moment, on 25 December 1761, the Empress died and was succeeded by Peter III, a zealous admirer of the king of Prussia in particular, and everything German in general. On his first day of rule, he ordered his forces to stop hostilities against Prussia and his treasury to pay retribution to the people of devastated Pomerania. Moreover, he did not stop there and, on 24 April, he signed a peace treaty with Prussia, giving up Russia's previous conquests and offering Russian troops to the service of Frederick the Great. Therefore, with the rise of a new emperor, yesterday's enemy became Russia's closest ally and yesterday's allies turned into enemies. Simultaneously, Peter III started preparations for a campaign against Denmark in order to change the borders of the Duchy of Holstein. If this war had occurred, it would probably have been useful for Prussian aims and useless to the Russian state, like everything else done by Peter III. Fortunately, the successful seizure of the throne by Catherine the Great put an end to his anti-national politics.

'I am quite bellicose,' said Catherine once to a foreign ambassador to the Russian court; 'however, I will never start a war without reason. If I am to start a war, then I will do it not like Empress Elizabeth, who did so to threaten others, but only when I find it convenient to myself.'

Russia's participation in the Seven Years' War cost more than 300,000 men and thirty million roubles, without securing any achievements or gains.

Similar examples of politics acting without any guiding idea can be found all through human history, though they do not always appear in a such clear form as with Peter III. The second type of mistake, when *politics is guided by a wrong idea*, occurs much less frequently. An example of this can be found in the politics of Russia from the beginning of the reign of Paul I to the death of Nicholas I. The main idea of this politics was *the struggle against revolution*.

The emperor Paul I, who laid the cornerstone of this policy, inherited from Catherine the Great a strong and formidable empire that instilled fear and respect in all the old powers of Europe. Under Catherine, a healthy state egoism defined all Russia's actions. Russian politics was strictly national—it had no other aim but the advantages and needs of the Russian people. Based on the country's fearsome military power, Russian diplomats let themselves adopt a very special tone in their relations with Europe. Echoing the language used by ancient Rome, they did not allow any rejection of the ultimatums they had drafted. Catherine had two main goals: the conquest of Poland and control over the exit from the Black Sea. She pursued these goals in a remarkably consistent way, intervening in the affairs of Western Europe only in so far as it assisted the achievement of her goals. Therefore, when the armies of the French Revolution flooded neighbouring countries with a wave of fresh ideas, Catherine, fully realising the spirit of her people, stayed absolutely calm. On the one hand, outraged by the rampant violence of the revolutionaries, she promised her help to the threatened monarchies and even mobilised Russia's military. On the other hand, it seems that she only wanted to use the troubles that then seized Western Europe to promote Russia's direct interests. [...]

This strong political inheritance was bequeathed to Paul I in November 1796 after the sudden death of Catherine. The new emperor, who was suspicious by nature and very jealous of his power, hated the very idea of the liberation movement occurring in the West. Therefore, regardless of the fact that foreign states were told that Russia had no intention of being directly involved in war with France, in his statement to the courts in Vienna, London and Berlin the emperor declared that 'he *feels the need to resist the outrageous French Republic* by all possible means, as it threatens the whole of Europe with the extinction of law, rights, property and good manners'. [...]

However, monarchic Europe was unable to stop the advance of the French Revolution by itself and asked help from Paul I. This is how the deposed king of Naples expressed himself in his letter to Paul: 'I offer Your Majesty, the most powerful and the only protector of the crowned kings, my sincere request: be my patron, supporter and avenger.'

Undoubtedly, from the perspective of its direct interests, Russia was indifferent to the future of Italy, whether it be in the hands of the kingdom of Naples or its successor, the Parthenopean Republic. However, flattered by the trust of the West European monarchs and believing that the Revolution should be suppressed before it reached the borders of the Russian Empire, Paul declared war on France. The aim of this war, as defined by the emperor to General Rimsky-Korsakov, was 'to stop the successes of the French Republic, to restrict its ability to disseminate the contagious ideas of dangerous liberty and restore the old thrones given by God to the monarchs'. Therefore, the guiding idea behind the war that began in 1799 was to fight against revolution in the name of the principle of monarchy.

Alexander I ascended the throne on 23 March 1801, after the assassination of his father. While, owing to his liberal education, he was inclined to sympathise with the principles declared by France, he thought that they should be promoted by the kings, who would bestow these freedoms from the heights of their thrones. He saw the violent behaviour of the revolutionaries as an abuse of power. Where the French saw an enlightened liberalism, he saw a crude mutiny. Driven by his views, Emperor Alexander wanted first to calm the storm created by the revolution and return France to its lawful borders. Only then did he propose to begin with 'the rearrangement of Europe, based on the rule of law and justice, ensuring the liberty and prosperity of its people'.

While the second (visionary) part of this programme remained unfulfilled, the first (realistic) one lay at the heart of Russia's foreign policy, igniting the wars of 1805 and 1807. Only after the meeting in Tilsit on 7 July 1807 did Alexander and Napoleon become closer. The French emperor clearly understood the significance of the alliance between France and Russia and tried to attract Alexander by all possible means. [...] The French desire to keep Russia on its side gave Alexander the opportunity to finally solve the Russian problem in the east. The long-cherished dream of Catherine came ever closer to being realised. Russia's actual interests did not clash with those of the French. The defeat of its neighbouring countries, Austria and Prussia, was very beneficial for Russia. Despite the temporary restrictions of the Continental System, the fierce war waged by Napoleon against England was also very beneficial to Russia, as England was one of Russia's traditional enemies. Unfortunately, this Franco-Russian alliance did not last. While there were many reasons for its failure, the main one that led to the final breakdown was the desire of Emperor Alexander to follow in the steps of his father and fulfil the role of 'the only protector of the crowned kings'. [...]

In other words, Alexander's desire to protect the lawful rights of monarchs destroyed this very beneficial alliance with France and led to the devastating invasion of 1812.

When the French and all their allies were forced to withdraw from Russia, Field Marshal Kutuzov proposed to end the war decisively, arguing that *'Napoleon does not represent any threat to Russia anymore and should be kept in shape to fight the English'*. Count Rumyantsev, the Chancellor of the Russian Empire, held very similar views. Even public opinion in Russia was against military operations abroad. However, despite all this, Emperor Alexander decided to be the saviour of Europe. Even after the defeat of the coalition in the 1813 Battle of Lützen, when Napoleon offered Alexander a separate peace agreement on very favourable terms, Alexander continued to pursue his goal.

STRATEGIYA

The Congress of Vienna proved the foolishness of Alexander's politics. It is well known that during this diplomatic meeting, England and Austria opposed Russia's annexation of the Duchy of Warsaw. Both nations feared that this would make Russia much too powerful, and signed a secret treaty against Russia on 22 December 1814. Only the surprising return of Napoleon from Elba forced them to abandon their plans, removing this hidden threat to Russia. When Napoleon found the text of this secret treaty in the cabinet of Louis XVIII at the Tuileries palace, he immediately sent it to Emperor Alexander, offering an alliance on any conditions. However, even after this, Alexander remained faithful to the coalition.

While the actions of Emperor Alexander were always marked by a degree of equivocation (even if on some occasions he could be very persistent), his successor, Emperor Nicholas I, had a very solid nature. Believing deeply in the divine origins of monarchy, he thought that the main task of government was to protect monarchy from any internal or external threats. 'Revolution is at Russia's gates,' he said during his accession speech, 'but I promise you that as long as I am alive, it will not get in.'

And indeed, when the July Revolution started in France in 1830 and the wave of troubles reached Belgium, Italy and certain parts of Germany, Emperor Nicholas I immediately mobilised his army and demanded the same of his allies Austria and Prussia. [...] 'We have to prove to the Jacobins that we are not afraid,' Nicholas used to say, 'and if we (monarchs) are destined to die, we will do so with honour, defending the breach with our chests.'

Austria and Prussia, however, were busy with their own internal affairs, and refused to intervene in the internal affairs of France. While the November uprising in Poland initially diverted Nicholas's attention as well, immediately after enforcing peace in Poland, the Russian government returned to its way of fighting revolution. Only this political idea can explain the Russian intervention in Hungary, as there was no other reason to justify this campaign. From the perspective of Russia's direct interests, the set of revolutions occurring in the Austrian Empire in 1848 was very favourable to Russia. A neighbouring country, whose unfriendliness Russia experienced on more than several occasions, was falling apart by itself without any efforts on the part of Russia. This could significantly ease Russia's efforts to solve its problems in the east.

In addition, the fall of the Habsburg dynasty would release various Slavic nations, which could easily fall under Russia's influence. Finally, Galicia, an old Russian region that was forgotten by Russia's cosmopolitan diplomats at the Congress of Vienna, could be reunited with Russia, allowing the country to acquire a strong natural border, the Carpathian Mountains.

While these were the real interests of the Russian people, Emperor Nicholas I sent his army to Hungary, declaring that 'the mutiny and lawlessness has spread across Austria and Prussia. Their madness and lack of control threaten our God-given Russia.'

The Habsburg dynasty was saved. However, the role of saviour of Europe, selflessly played by Russia's government for so many years, gained Russia not only the sympathy of several monarchies that were saved, but also the antipathy of many nations that got used to seeing Russia as the main enemy of their development as free countries. The Crimean War was the atoning sacrifice for this mistake.

In analysing this period of more than half a century, it becomes clear that all wars waged during this time by Russia in the west were driven by one main idea—*to fight against revolution in the name of the principle of monarchy*. While Russia's politics was distinctively consistent, its guiding idea was wrong, because, regardless of the relative success in all these wars, Russia derived absolutely no benefits from them. The opposite is true. By participating in these wars, Russia made the achievement of its historical aims even more difficult, as it helped to create the power of the states that currently stand in Russia's way. Moreover, the campaigns of 1799, 1805, 1807, 1812, 1813, 1814 and 1848 and the Crimean War in 1853–56 cost Russia approximately two million men and almost a billion roubles, and this is without considering the indirect losses to the national economy.

Regardless of all these sacrifices, the spirit of the time could not be stopped. Political and social reforms were gradually carried out in all European states. Even Russia itself, after the Crimean War, entered the path of transformation.

II. Politics must prepare a favourable situation for strategy's actions

Politics has several means, allowing it to prepare a favourable situation for strategy: the making of alliances, the securing of neutrality, subversion of an enemy's population, financial preparations, and the creation of positive public opinion.

1. *Making alliances.* On the one hand, any military assistance will always be welcomed. Therefore, it is not advisable to automatically disregard alliances even with the weakest allies. On the other hand, different alliances have a very unequal price from the viewpoint of strategy. Politics should not be satisfied with accidental allies. Even during peacetime, it should identify potential future enemies and simultaneously search for useful allies. Therefore, politics should carefully measure the value of a potential ally for strategy. This values consists of several aspects.

   The first condition is, obviously, *the strength of the ally.* However, this strength should not be defined solely by the size of the population and the existing military system. The strength of a state is an outcome of many very complex variables, such as its governmental and social structures, its economy and national wealth, the qualities of its culture, the character of its people, the personality of the head of state, etc. Each of these variables has its own influence on national military strength. For example, sometimes a state with a very strong military is unable to wage a prolonged war, because its finances are completely disorganised. In another case, a state with a strong military and a good economy cannot guarantee a stable alliance, because its decision-making process is either a toy in the hands of political parties or its sovereign is weak and indecisive. There are also states that are strong only in times of prosperity and fall apart at the first serious blow of the enemy, because they were created as the artificial union of too many different peoples.

   Moreover, in assessing the strength of an ally, it is important to bear in mind that the extent of its participation in war depends on the degree of its *interest* in doing so. While looking for useful allies, politics should, first and foremost, search for them among states that share mutual interests with us. However, if there are no such interests, politics should create them, attracting allies by promising real benefits. Only alliances signed where there are mutual interests can be strong. Otherwise, immediately after the first setback, the ally will seek peace with the enemy or will even cross sides. This is usually the fate of all alliances based on feelings of gratitude, the pursuit of secondary aims, or personal sympathies. The best example was the Crimean War, when Emperor Nicholas I believed that Austria and Prussia would become his allies or, at least, would stay neutral. His belief was based on the assumption that all three countries were linked by the Holy Alliance, a feeling of gratitude for help provided on multiple occasions, and friendship between the monarchs. These conditions, however, were not enough. Since the actual interests of Austria and Prussia contradicted those of Russia, they decided to take a position against Russia. Especially instructive in this respect was the behaviour of Austria, which had been saved by Russia only few years beforehand. In January 1854, Emperor Nicholas, who was already at war with Turkey and was threatened by English and French fleets, which entered the Black Sea, sent Adjutant General Prince Orlov to Vienna with the modest request that the Austrians stay neutral. Austria, however, put forward a whole set of demands: that the Russians not cross the Danube, withdraw from Moldavia and Wallachia after the end of the war, and generally avoid the destruction of the existing order in Turkey. It has been said that when Orlov, outraged by this lack of gratitude, reminded Emperor Franz Joseph I of the help given by Russia in 1849, he calmly replied: 'There is no role for emotions in politics, only benefits.'

Sometimes, a powerful state might have weaknesses in some particular respects. In that case, it will appreciate more the ability of its ally to make up for what it lacks, rather than its overall strength as discussed above. For example, when starting the war against Napoleon in 1805, Austria and Russia had significant military power, but their financial capabilities were not in a good shape. Therefore, they signed an alliance with wealthy England, which could not offer a significant army but was able to finance the mobilisation and deployment of the armies of its allies. [...]

The second condition that defines the value of an ally from the point of view of strategy is its *geographical position*. This defines the place of the initial concentration of force, giving an advantage (or not) to the starting point of the main operations. From this perspective, one can divide all possible cases that might occur in practice into four main groups:

STRATEGIYA

a) *The ally is beside us.* This position allows for the immediate deployment of both armies on one common line and the simultaneous opening of the campaign. In addition to this, there are several other positive advantages in the close location between allies. Their bases of operation become wider, to the extent that there are certain advantages in uniting them into one shared base of operations. Moreover, allies based in the same position experience the same level of danger and, therefore, are forced to wage war with the same degree of determination. These advantages are so significant that they turn this geographical position of the allies into the most favourable one. For example, Germany and Austria are positioned in this way when they face Russia.

b) *The ally is between us and the enemy.* In this situation the allies will have to sacrifice either time or the option of mutual action. In the first case, in an attempt to unite efforts, our ally will have to wait for the arrival of our army. In the second case, if the two do not want to waste time, our ally will have to start action before the arrival of our army. While this position has certain disadvantages from the perspective of our interests, there is also one main advantage. If there is an unfortunate development, the theatre of war is not on our territory, but on the territory of our ally. Therefore, we have an opportunity to seek peace with the enemy on favourable conditions before the fighting reaches our country. This was the situation with Russia in the wars of 1805 and 1807. In both cases Austria and Prussia were unwilling to waste time and wait for the arrival of the Russian Army, and thus they were defeated. When the subsequent struggle proved the absolute superiority of Napoleon, leading to the peace treaties of Pressburg (1805) and Tilsit (1807), it turned out that Russia did not lose much. On the contrary, according to the Treaties of Tilsit, Russia got from its former ally the Białystok Department.

c) *The ally is placed behind us.* This position represents all the disadvantages of the previous example, without offering any advantages. In this case we play the role of the victim when our ally leaves it to us to fight his war at our peril. The example of

Austria and Prussia proves the possible detrimental consequences of this situation. While Russia finished the disastrous wars of 1805 and 1807 unscathed, and even won some territorial gains, Austria, according to the Treaty of Pressburg, had to pay reparations of 100 million guilders and give up territories with three million inhabitants; and Prussia, according to the Treaty of Tilsit, lost more than half of its territory.

d) *The ally, like us, has a border with the enemy, but it is separated from us by a neutral state, or by sea, or by the enemy's territory.* The main disadvantage of this situation is that the allied armies can only be united on the enemy's territory. Therefore, an enemy placed between allies can prevent the allies from uniting their armies, by concentrating its forces on one ally, while delaying the attack of the other by means of a relatively small force. This theoretically simple idea lay at the heart of all operations conducted by Frederick the Great during the Seven Years' War. While this position has many disadvantages for the allies, in certain situations it can offer some advantages too. On the one hand, a geographically close but weak ally will not contribute much to his stronger partner. On the other hand, if this weak ally is based on the enemy's flank, and is separated from us by a sea controlled by our fleet, then its territory can serve as a great base for offensive operations against the enemy's communication lines. This role could have been played by Denmark during the war of 1870–71 only if France had supplied enough soldiers and means of transport to establish powerful landing forces there.

While attracting the most useful allies to its side, politics also should simultaneously destroy the alliances of the enemy. To deny an enemy allies, politics can even use open force, especially when these potential allies are weak and unable to withstand prolonged pressure. This is exactly what Prussia did before the spring of 1866. Anticipating that Hanover, Saxony and Hesse might join Austria, Otto von Bismarck decided to paralyse these countries in advance. He demanded that the governments of these countries demobilise their militaries and reduce them to a peacetime level. When they refused, Prussian forces invaded these countries, thereby securing their flanks and rear for the main offensive against Austria.

2. *Securing neutrality*. In the case where an attempt to bring a certain country into alliance with you has failed, it is important to guarantee that it will not take the side of the enemy. A state of neutrality with a neighbour is very important for strategy, as it allows one to concentrate all one's forces against the enemy. The best example of this was the Franco-Prussian War. Thanks to the sympathetic neutrality of Russia, which also helped to keep Austria neutral, Prussia did not have to keep more than one division in its rear and could invest the whole might of its army against France.

   Another example was Russia during the Crimean War, when it found itself in a completely opposite situation. Waging open war against Turkey, France, England and Sardinia, Russia failed to secure the neutrality of Austria and Prussia, which adopted a very hostile attitude to Russia's position. On the one hand, during this war Russia had an army more than a million strong. On the other hand, owing to the need to keep significant forces to secure its borders against a possible escalation, Russia failed to deploy in Crimea enough forces to win against the hundred-thousand-strong enemy force. This was a consequence of the unsatisfactory political preparations for the war.

   It is important to note that one would be wise to count on the neutrality of a neighbouring country only if it is capable of protecting this neutrality. Otherwise, it might happen that the enemy marches its forces across the neutral state, regardless of all international laws. Napoleon I used to do so sometimes. Belgium and Switzerland may face a similar fate in the case of future war between Germany and France. This is probably why these countries, not satisfied by Europe's ceremonial promises of neutrality, decided to increase their military power—Belgium has turned Antwerp into huge fortress-city, and Switzerland has been organising a vast militia. [...]
3. *Subversion of the enemy's population*. Almost every country carries within it the germ of internal political or social disease. Therefore, in addition to its external threat, it might have internal ones that subvert its power. Good politics should find them by scrupulously studying the enemy's governmental structure, social life, and the ruling classes of its people. Only a close familiarity with the internal

life of the enemy will help to discover its weaker sides, which are usually hidden by the superficial gleam of its external might. In exploiting these weaknesses of the enemy's state, good politics will always find collaborators who are dissatisfied with the existing order. Generally speaking, there are four possible developments:

a) *The population of a whole region in an enemy state takes our side.* This situation may occur only when there are peoples inside the enemy state that are seeking national independence or willing to unite with us. All Russia's wars against Turkey were fought under such a situation.

   Peter the Great, starting his struggle against the Sublime Porte, observed that almost the whole Balkan Peninsula was inhabited by peoples who shared either the same faith with Russia or the same ethnic roots. Therefore, Peter's main plan was based on his attempt to have these peoples rise up against the Ottoman government. These ideas were very successfully promoted by Russia's ambassador to Turkey, Count Tolstoy. The Orthodox Christians of Turkey openly promised to help Russia's efforts. Before starting the campaign, Peter issued a call, inviting them 'to rise up for the faith and fatherland, and free themselves from the non-Christian yoke'. Unfortunately, the defeat in the 1710–11 Pruth River Campaign did not allow Peter to achieve his ambitious plan. However, his idea of shaping the struggle against the Turks along national and religious lines did not die with him. It was successfully embraced by his successor and lay at the heart of all Russia's wars against Turkey.

b) *A certain class of the enemy's population takes our side.* This can occur only in states undermined from the inside by social diseases. The harsh material conditions and lawlessness of the lowest classes can push them into the hands of the enemy, making them choose the immediate benefits of their class instead of the interests of their state.

   When invading neighbouring countries, the army of the First Republic had to fight only against their governments and aristocracy. The middle and lower classes, however, greeted the French with gratitude, as they brought with them long-desired freedoms and, more importantly, social reforms. The sympathy of the local masses was one of the main reasons for France's quick success. Napoleon I also used this powerful factor. While building absolutism in France and reviving its aristocracy, Napoleon I continued to promote the principles of the Revolution during his campaigns, bringing new values to the peoples conquered by his arms. His conquests were, in fact, the main reason for the resurgence in many neighbouring countries: Italy, Switzerland, Netherlands, some parts of Germany, and even backward Spain were forced by his actions to carry out many liberal reforms.

c) *A certain political or religious party takes our side.* This can occur in a state that does not have a solid, well-established order. For example, such a situation of religious instability was found by Gustavus Adolphus in Germany during the Thirty Years' War. Waging war against Emperor Ferdinand II and other members of the Catholic League, he decided to lean on the Protestant parties. Another example was the politics of Catherine the Great. While seeking to annex Poland, she used the political rivalry inside the country, trying to find assistance from different political parties fighting for power.

d) *Our goal is supported only by certain individuals.* While finding individual supporters in the enemy's camp is always possible, it becomes especially easy when moral decay and the general pursuit of material wealth occupy the society. This social illness usually begins with the upper classes, and only then spreads across the whole body of the society. The idea of the fatherland loses its unifying power and the sense of patriotism weakens, being replaced by individualistic interests and personal advantages. Any country in such a state of decay is easy prey for its neighbours, regardless of its visible external might, huge army, enormous fleet or highly developed culture. Even the greatest nations can find themselves in this situation when they are in their period of decline, i.e., in the final stage of their existence. The final years of ancient Greece and Rome are good examples of that.

4. *Financial Preparations.* On the one hand, Frederick the Great used to say that 'war requires three things—firstly, it requires money;

secondly, it requires money; and, thirdly, it requires money again'. On the other hand, he found it possible to fight for many years against the enormous European coalition with relatively limited financial means. [...] In fact, history shows that during the days of cabinet wars, governments did not feel too restricted by the question of finance in carrying out their foreign policy. In a worst-case scenario, when a country was failing to attract creditors, the money required could be gathered by other means, such as by introducing new taxes or increasing old ones, by enforcing special contributions from subjects or taking compulsory loans from merchants, by debasing the coinage, etc. [...] Even in the time of the French Revolution and Napoleon I, the financial situation did not have any significant influence on the capacity of the state to wage wars. The Bourbons left the country on the brink of economic catastrophe, but it managed to wage several victorious wars and, after a twenty-year period of war, it was still a prosperous country with developed industries and well-maintained finances. [...]

Nowadays, however, finances have a significantly greater influence on the capability to wage war. On the one hand, the wars of 1812, 1813 and 1814 cost Russia 157,450,710 silver roubles, or, on average, 52,483,570 each year. On the other hand, the devastating Crimean War, which lasted two and a half years, cost Russia 796,770,000 silver roubles, or 318,708,000 each year. Russia's expenditure during the last war against Turkey, which lasted approximately a year, was 1,020,578,489 silver roubles, despite Turkey being a second-rank enemy.

Therefore, there has been a ten-fold increase in the costs of war during the last century. Even if we take into consideration inflation, the increase in the costs of war has been significantly higher than the inflation rate. Moreover, while war has always had a devastating impact on the national economy, nowadays, owing to the greater development of trade and industry, this impact is more destructive than ever before. These enormous direct and indirect costs of contemporary wars emphasise the increasing importance of financial preparation for war.

Nowadays, there is only one way to secure the funds required by war—credit arrangements. In contemporary wars, a country's military budget, regardless of how rich this country is, can satisfy the requirements of war only in its early stages. The funding of any subsequent fighting will require the government to borrow large amounts of money. Therefore, one of the most important tasks of politics is to prepare the ground for borrowing at home and, simultaneously, to subvert the adversary's credit-worthiness. In an attempt to achieve this, politics should fully exploit the other nation's interests and sympathies, so as to facilitate any financial negotiations.

Colmar Freiherr von der Goltz[2] argues that 'the contemporary way of war is impossible without an ability to generate significant credit. On the other hand, it is possible to argue that a state cannot be defeated as long as it has money and credit. No great power will suffer from a shortage of men capable of bearing arms as long as it has sufficient material means. Only those who can wage a long war have the potential to achieve victory.'[3]

5. *Creation of positive public opinion.* While public opinion obviously has no direct influence on strategy, it exerts quite significant pressure on the decision-making of the government, especially in constitutional states. This pressure can be so significant that a government which previously was inclined to stay neutral will be forced to change its position and join one of the active belligerents. [...]

The increasing role of public opinion pressures politics to avoid anything that can cause people to rise up against it, even if this politics solely pursues national interests. While having mercilessly self-serving foundations, politics should endow its actions with a lawful appearance. Therefore, there is no better way of creating a reason for going to war in future than to insert it into a peace treaty.

One of the best examples of such forethought was demonstrated by Catherine the Great. While the guiding aim of her eastern politics was the destruction of Turkey and control over access to the Black Sea, Catherine well realised that this enormous task could

---

[2] Wilhelm Leopold Colmar Freiherr von der Goltz (1843–1916) was a Prussian field marshal and military writer.

[3] The exact source of this quote is unknown. A similar idea can be found in Freiherr von der Goltz, *The Nation in Arms*, translated by Philip A. Ashworth (London: W.H. Allen and Co., 1887), p. 122.

not be achieved by one blow and would require a whole series of wars. Therefore, negotiating the Treaty of Küçük Kaynarca in 1774, she decided to insert an article that would always offer Russia a good reason for re-starting the war. According to article seven of this treaty, the Sublime Porte recognised Russia's right to protect Russian Orthodox Church practices across the Ottoman Empire. The unusual ambiguity of this article and the easily combustible outbursts of Islamic fanaticism, against which the Turkish government was powerless, opened a large window of opportunity for Russia to intervene in the internal affairs of the Ottoman Empire. In this way, the noble humanitarian mission of protecting Russia's co-religionists lay at the heart of all Russia's wars against Turkey. This article was passed from one treaty to another, slowly acquiring more and more power and providing a justifiable and lawful reason for starting the next war. This adventurous situation was ruined by the Treaty of Paris in 1856, which put Christian subjects of the Ottoman Empire under the common protection of all the great powers.

Nowadays, the press is the main weapon for attracting public opinion, and Bismarck understood this before anybody else. He attracted many talented journalists to the civil service, making them channels of communication between him and the press. In this way the press were fed and directed in line with his programme. Every diplomatic step was carefully prepared beforehand by agitation, and so Bismarck's ideas were accepted by the German masses in advance. Moreover, during international crises, he always tried to attract international public opinion to Germany's side and shed negative light on its adversaries.

III. Politics must choose the right moment to start war

The best moment to begin a war is probably when our adversary is not ready for it and the countries that support him are busy with their own internal affairs. Sometimes, the natural development of history, without any intervention from our side, can create such a favourable moment for us. At all other times, this moment should by created by the skilful efforts of our diplomacy.

STRATEGIYA

In any case, by using this moment, politics not only eases the task of strategy, but also creates an opportunity to win the best possible benefits from military action and, in a case of failure, to minimise the losses.

One of the best examples of a skilfully chosen moment for war was Bismarck's decision to start the Austro-Prussian War. Firstly, Austria was still recovering from its unsuccessful campaign in Italy in 1859. Its military system was outdated and its officer corps, equipment, recruitment and training were undergoing significant reforms. Moreover, the national struggle inside the country was at its peak, with Hungary almost openly fighting for its independence. Secondly, Austrian's allies—Saxony, Hanover, Hesse-Cassel, Württemberg, Bavaria and Hesse-Darmstadt—were even less ready for war than Austria itself. Meanwhile, Prussia had a well-maintained army and was able to rely on Italy, with whom an alliance had already been negotiated in April 1866. [...]

Our last war against Turkey, however, represents a completely opposite example. Since the times of Catherine the Great, Russia had had several opportunities to resolve its eastern question or, at least, make significant steps towards its resolution. Such convenient moments occurred during Russia's alliance with Napoleon I, during the Hungarian Revolution in 1849 and during the Franco-Prussian War in 1870–71.

Russia, however, failed to use these favourable windows of opportunity, usually starting its wars against Turkey when the attention of Europe was not distracted by something else. Moreover, under the pressure of Russia's idealistic politics towards the West, its views on the eastern question had also changed. Russia's healthy national egoism was increasingly replaced by undefined humanitarian impulses. For Empress Catherine, control over the Black Sea Straits was *the aim*, and the support of the Balkan Christians was *a means* to achieve this aim. For her successors, however, this means somehow turned into the aim. While Catherine exploited the sympathy among Christians to the benefit of Russia's national interests, the politics of later governments selflessly sacrificed the blood and money of the Russian people. Instead of promoting national interests, they tried to assist the co-religious peoples of Greece, Bulgaria, Serbia and elsewhere who were allegedly loyal to Russia. [...]

Therefore, because Russia has constantly failed to choose the right moment for its wars against Turkey, the politics of Russian governments introduced various restrictive factors which not only made the implementation of strategy more difficult, but also rejected in advance any opportunity of benefiting from future military achievements.

IV. Politics must define the political aim of war and then offer strategy full freedom of action

The political aim of any war has to be considered in the context of the political idea that guides the country in a specific period of its history. Therefore, it is important that the chosen political aim should be as big as possible, allowing the country to make as big an advance towards the national idea as possible. Politics should carefully consider what it can achieve in a historical moment with the available means. Any mistake in this consideration, such as the definition of a political aim that is too wide, may lead to a situation where waging war is beyond the power of strategy and it would fall apart from the unbearable pressures.

In addition to the political aim, which differs from war to war, all wars have the same military aim—to break the enemy's resistance and compel him to submit to our will. Since choosing the way of action to secure this second aim (i.e., drafting the war plan and executing it) is the responsibility of strategy, politics should not interfere in this professional sphere.

Our last war against Turkey demonstrates best the detrimental consequences of a situation where politics interferes in strategy. As we have already seen, the political aim of this war was to compel the Sublime Porte to give certain rights to its Christian subjects. However, politics did not limit itself to this political aim and, interfering in the sphere of strategy, defined also the military aim of this war—occupation of Bulgaria and the Balkans. Many high-level officers tried to oppose this aim, claiming that occupation of Bulgaria and the Balkans would not force the Sublime Porte to submit to our will, as this could be achieved only by the complete destruction of the Turkish military and the subsequent occupation of Constantinople. One of our most talented generals argued: 'It is absolutely necessary to achieve full military control over Constantinople and the Bosporus. The only thing that should stop us from pursuing this goal is an offer of peace by Europe and the Sublime Porte on the very same conditions as if we were already inside Constantinople. Without this offer, any delay on our way to Constantinople will lead to *our destruction*. Therefore, diplomacy should not *confuse* us in any way, *interfering* in our affairs with any kinds of compromises.' However, our diplomacy, tied by certain obligations, continued with its direction of the action, and Russia started the campaign with the military aim of occupying Bulgaria.

When the inevitable course of events forced the main part of Russia's army (regardless of promises previously given) to cross the Balkans, the commander-in-chief wanted to immediately occupy Gallipoli and Constantinople, i.e., to control the Dardanelles and the Bosporus. While the Turkish ambassadors, who arrived at the Russian headquarters at the beginning of January 1878 in an attempt to negotiate peace, had no objections to this move, it was Russia's diplomacy itself that (by observing restrictive obligations that had been voluntarily undertaken) did not allow Grand Duke Nicholas to accomplish his considerable plan. This indecisiveness of Russia's leadership was immediately noticed by the English, who sent their fleet of battleships into the Sea of Marmara.

Therefore, the situation of the victorious Russian Army which had stopped at the walls of Constantinople became very critical. It was stationed in the middle of a devastated land and all its supplies were shipped by sea from Odessa to Constantinople. Meanwhile, in addition to the powerful Turkish fleet, a fleet of English battleships controlled the Bosporus Straits. Since Russia had no military fleet to protect its lines of communication, the enemy could easily destroy them in case of any escalation. Without the ability to supply the army by sea, Russia would have to use very long land lines through the whole Balkan Peninsula, along unpaved roads and across several natural obstacles such as the Balkan Mountains and the Danube. Moreover, these lines would pass by the Turkish fortresses of Shumen and Varna and, as if this was not enough, they would be exposed to possible strikes from the almost openly hostile Austria and Romania. Even telegraph communications between the army headquarters and St Petersburg went through Constantinople, and sometimes the Turks delayed the transmission of encrypted telegrams which they considered especially important.

Under such circumstances, as a result of its own politics, Russia had no choice but to yield at the Congress of Berlin. It appeared in the court of Europe not as a fearsome victor, but as a power that had carelessly fallen into a strategic trap.

However, if Russia's strategy had been free from politics and managed to control the straits, the situation would have been entirely different. In this case, all communications by the Black Sea would be completely secure; the defence of the whole Black Sea coast would be reduced to the defence of one point (the Dardanelles and the Bosporus); the forces assigned to protect the long coast would be freed, as well as the forces left to protect land communication lines. In this situation, any attempt to force Russia out of its strong positions on the Balkan Peninsula would require at least a half-million-strong army, which England did not have and Austria would not send, fearing to expose its borders. [...]

V. Politics must give strategy the required means

Both internal and foreign politics are equally responsible for fulfilling this obligation. The *internal* politics defines the size of the military and the defence budget. In constitutional states these two are defined through a process of intense bickering between different political parties and the subsequent series of compromises enforced on the government. Under such circumstances, it is not always easy to establish the military power required to fulfil defined political aims. For example, Bismarck experienced this first-hand during his first years of power. In 1862, the newly appointed Minister President defined the guiding aim of his politics—the unification of Germany under the rule of Prussia. Achievement of this goal required a strong army, but the assembly, controlled by liberal representatives, was opposed to an increase in Prussia's defence budget. On the one hand, it was possible that they would give their approval if they knew that the money would be used to resolve a long-standing national issue. On the other hand, Bismarck could not fully disclose his cards and was forced to limit himself to allusions only. 'It is not by speeches and majority resolutions that the great questions of the time are decided, but by iron and blood,' claimed Bismarck during one of the meetings of the budget commission. 'Prussia has to coalesce and concentrate its power for the opportune moment, which has already been missed several times; Prussia's borders are not favourable for a healthy, vital state.'[4]

However, this explanation was not enough for the liberal representatives, who were afraid to give a powerful army to the former leader of the conservative party, fearing that it could be used to enforce various reactionary measures. Realising that the budget deadlock could not be resolved in a constitutional manner, Bismarck decided to bypass the legislature. [...] This conflict between Bismarck and the liberals lasted for many years, and even the successful war against Denmark did not solve it. Only the victorious war against Austria brought Bismarck the necessary political support. On 5 August 1866, William I, the king of Prussia, talking at the Landtag, pointed to the great achievements of Prussian politics and proposed that internal conflicts be resolved. Following his speech, the majority of the representatives, 230 against 75, approved all expenditures of the government.

On the one hand, in absolutism the same questions regarding the size of the army and the defence budget can be solved without any strident opposition (this is why the military systems of these countries can enjoy an incomparably greater development). On the other hand, absolute power can lead to the opposite extreme to the situation in a constitutional state.

It is important to remember that in every state, within certain conditions, the military has a certain desirable and necessary size, failure to achieve which will have serious consequences. Moreover, the military power of a state does not depend only on the size of its army, but it also requires well-managed financial support, developed industry, lines of communication, etc. Finally, the very military itself is an exact image of the people that created it. The state's organisation, the economic structure, the relations between classes, the legal system and even the main guiding political idea—all these are reflected in the organisation and training of the military. The general aspects of a nation's life not

---

[4] Martynov refers to Bismarck's famous speech titled 'Iron and Blood'; see Otto von Bismarck, *Reden 1847–1869* [Speeches, 1847–1869], edited by Wilhelm Schüßler, vol. 10 of *Bismarck: Gesammelte Werke* [Bismarck: Collected works], edited by Hermann von Petersdorff (Berlin: Otto Stolberg, 1924–35), pp. 139–40.

only influence the preparation of the military during peacetime, but also shape the specific characteristics of the nation's art of war. In times of war, a nation mobilises all its strengths, physical and spiritual, utilising all available technological innovations, all scientific inventions and all possible moral impulses. Therefore, the very direction of military operations (the most difficult task that can be given to a human being) is in the hands of people who grew up in the environment of a particular nation, moulded by its intelligence and education.

Owing to the close dependence of military affairs and the art of war on other parts of human life, these aspects cannot be ignored when thinking about military power. A state's military power is strong and sustainable only when the development of the armed forces does not contradict the general health of the state and all parts of national activity develop with a certain degree of harmony. History proves that any violation of this basic principle leads to very detrimental consequences.

The best example of this violation occurred during the times of Emperor Nicholas I. It is a well-known fact that Nicholas's sole attention was given to his military. The interests of Nicholas's army pushed all other needs of the country into the background. However, while the Russian military of the 1850s was considered by many as one of the exemplary armies in Europe, a few foresighted people doubted Russia's military power, comparing it to a colossus with feet of clay. Unfortunately, this opinion of the few turned out to be true when 60,000 Allied troops landed in Crimea in September 1854. While Russia had a colossal army of 1,335,000 men, it failed to concentrate them to face the enemy in time. This happened not only because many units were dispersed across Russia's long borders, but also, and more importantly, because Russia lacked a well-developed transportation system. Moreover, Russia's military equipment was outdated in comparison with that of European militaries—the consequence of weak technological development in Russia. Russia's officer corps did not match the requirements of contemporary war—the consequence of the uneducated social environment in which the officers grew up. The military was lacking initiative—the consequence of the fact that individual independence was generally suppressed in Russian society. [...] The military supply system was plagued with corruption—the consequence of the low intellectual and moral levels of the society of that time combined with the lack of a free press.

Therefore, despite the enormous amount of money spent on the military and regardless of its heroism, Russia's army could not withstand the outcomes of war. The colossus of military power, built by Emperor Nicholas I with so much care, collapsed at the first challenge because it was built on sand.

A large army, built and maintained during peacetime, does not ensure that military strategy will receive all the means required by war. *Foreign politics* can choose so unfavourable a moment for war that the majority of these forces may be assigned to the protection of borders and operational lines. Regardless of the situation, however, all available forces should be sent into the theatre of war without any hesitation. Numerical superiority will not only ease the task of strategy, but also lighten the burden of war for the state.

A campaign started from the beginning with all force will end sooner and will be cheaper. Moreover, its impact will be stronger and other states will have no time to intervene and reduce the benefits of victory. While there are possible cases (mountain or desert wars, or expeditions against barbaric tribes) in which the size of the army can be limited, this is not the case for wars in Europe. The general rule of big European wars is that the politicians should not be afraid of sending too many troops to battle and the only thing that they should be scared of is sending too few. Unfortunately, the whole history of Russia is marked by violations of this rule. While Russia's internal politics have been constantly busy creating a big army, its foreign politics repeatedly left strategy with insufficient means to match either the enemy or the importance of the defined aim. While this can be explained by an unsuccessful choice of the moment of war or by a failure to conduct all necessary preparations, the lack of attention by politics to the requirements of strategy is probably a better explanation. [...]

VI. Politics must skilfully exploit the results achieved by strategy

In fulfilling this last responsibility, politics should fight not so much against the enemy, whose resistance will usually be broken by the power of force, but against the intervention of neighbouring countries. Regardless of the degree of success, a war always temporarily weakens the state. A war stresses all the powers of the nation and is always followed by a period of material and moral exhaustion. This period of temporary weakness can be exploited by neighbouring states, which, on the pretext of mediation, may intervene in the negotiations for peace, stealing from the victor the fruits of his successes. Many successful wars were nullified in this way, bringing to the victors nothing more than military glory and material losses.

An ability to extract the biggest possible benefits from the successes achieved by strategy: this is the ultimate crown of diplomatic art, without which all preparatory efforts lose their meaning. Taking into consideration the complexity of this art, it is important to elaborate on one of its most important and relevant aspects. In negotiating peace, politics should have in mind the next possible war and create, in general terms, the most favourable situation for starting it. In this way, when politics completes its obligations to strategy in the previous war, it simultaneously begins influencing the strategy of the next one.

In 1871, during the peace negotiations with the French government, Bismarck closely followed Moltke's advice on the issue of the new border. The German strategists insisted that Metz fortress be given to Germany together with the Duchy of Lorraine. For a very long time Adolphe Thiers did not accept this request, offering different possible and seemingly favourable alternatives, some of which found many supporters among the Germans. Moltke, however, continued to insist, and Bismarck felt obliged to support him. 'I cannot agree on any replacement for Metz,' he said to Thiers, 'because Metz is irreplaceable from the military point of view.'

The result of such close attention by politics to the needs of strategy was a new Franco-German border that nullified both potential French defence lines (the Moselle river and the Vosges mountains). While in a case of war the line of the new border would force the French base of operations all the way back to Paris, the Germans were safe from invasion by their enemy behind the strong defence line on the Rhine and could use the newly obtained territories as a great bridgehead for offensive operations into France. The quadrangle of fortresses comprising Koblenz, Mainz, Strasbourg and Metz penetrated deep into French territory, placing its spearhead of Metz only a hundred and eighty miles from Paris. Therefore, building on Moltke's advice, Bismarck created a very favourable starting situation for a future war.

STRATEGIYA

In 1875, it seemed that the right moment to use this situation had arrived. France lay open and unprotected in front of the German forces. Bismarck's plans, however, were suddenly destroyed by Russia, which came to its senses and stopped the blow raised above France's head. Saved from their final destruction, the French started feverishly to fortify their eastern border. Although in the matter of a few years they successfully fortified the whole area between Switzerland and Belgium by two lines of interconnected fortresses and bastions, this enormous work also required similarly enormous investment.

After this positive case of Bismarck's politics, it is also important to analyse a negative example. Even a simple glance at the map of Europe reveals that the natural border of Russia with Germany and Austria should follow the Vistula from the Baltic Sea to the San River, then follow the San River to the Carpathian Mountains and all way to the Prut River. This border of natural obstacles would not only offer Russia many strategic benefits, it would also coincide with the ethnic border of the Russian people. Even Napoleon I recognised the Vistula as the true border of Russia, and, during the short period of alliance between France and Russia, he made more than several offers to Russia's diplomats in this direction. After the war of 1812, Count Rumyantsev and Field Marshal Kutuzov held a similar opinion, suggesting that once the enemy was forced out of Russia, it should fortify itself on the Vistula and not be so concerned about saving the rest of Europe. Emperor Alexander, however, insisted on pursuing his cosmopolitan beliefs. At the Congress of Vienna, Russia even returned to Austria the Tarnopol district (in eastern Galicia) which it had received from Napoleon I, getting in return the Duchy of Warsaw. During the following years Russia could have fixed this mistake on several occasions and traded the Polish provinces for the eastern bank of the Vistula, eastern Prussia and eastern Galicia. In 1849, 1866 and 1870, Russia would not even have had to fight to get them. Austria and Germany would have happily accepted this swap in exchange for Russia's neutrality. Russian politics, however, missed this excellent opportunity. [...]

Conclusions

This work has focused on a series of obligations that politics should fulfil if it chooses to use its most decisive weapon—war. On the one hand, it becomes clear that politics should clearly understand the needs of strategy. On the other hand, strategy is also required to support politics by force. Therefore, knowing the plans of diplomacy is not enough; strategy must also be aware of the specifics of the international situation at each given moment.

Consequently, politics and strategy are linked together, and their very tight connection should not be interrupted even for a minute. The best way to achieve this is to unite both in the hands of an ingenious sovereign, such as Alexander the Great, Frederick the Great, Napoleon I and others. However, such unification occurs very rarely. Usually we see that these two professions are divided. Therefore, if, by any chance, one nation simultaneously acquired a great diplomat and a great strategist (and only if both decided to work together towards one common goal), then it would achieve in a few decades things that otherwise would require centuries. The best example of this tight cooperation was demonstrated by Bismarck and Moltke, who were greatly assisted by the German Minister of War, Albrecht von Roon. The connection between these three was best expressed by William I, who proclaimed during a dinner after the victory against the French forces: 'You, General Roon, honed our sword. You, General Moltke, directed it. And you, Graf Bismarck, raised Prussia by your politics during the last several years to its current height.'

Our only hope is that in anticipation of practical outcomes from the forthcoming peace conference in The Hague,[5] similarly strong connections will be established between Russia's diplomatic and military departments. Any successes in foreign politics can occur only when this connection exists. And Russia, more than any other great power, can expect these successes because of its governance structure. Since politics and strategy equally require secrecy and decisiveness, they have a better chance of succeeding in autocratic countries. If Peter the Great had been a constitutional monarch, he would have failed to achieve even a tenth of what he successfully accomplished in reality. [...]

---

[5] Martynov refers to the Hague Convention of 1899.

3

THE FOUNDATIONS OF STRATEGY[1]

Nikolai Petrovich Mikhnevich

I

The Characteristics of Contemporary War

<u>The Conditions of Life of Contemporary Peoples</u>

The life of contemporary civilised peoples is characterised:

1) *in its material dimension*: by the widespread use of the steam engine and electricity, the introduction of aviation, and a general prevalence of technology in all spheres of life.
2) *in its economic dimension*: by the capitalistic (timocratic or property-based) structure of the state's affairs, widespread development of manufacturing and processing industries, and the presence of a significant part of the population (up to 70 per cent) living in the cities, working in factories and plants.

---

[1] This chapter is based on the translation of selected chapters from Nikolai Mikhnevich, *Osnovy strategii* [Foundations of strategy], (St Petersburg: Tipografiya Trenke i Fyusno, 1913).

3) *in its spiritual dimension*: by a keenness to pursue the idea of *nationalism*—a desire to unite large national groups.

In defining these characteristics, it is important to remember that they do not characterise the life of all peoples in the world. For example, even Russia does not fully answer them, as it still has not fully passed beyond *the natural stage of civilisation*. A large part of the Russian people's wealth still depends on the production of goods from nature (bread, livestock and unprocessed materials), and the vast majority of the people (85 per cent) still live by husbandry, hunting, fishing, etc.

Also, we should not forget that we still find some barbarian tribes living in a number of distant and wild places in the world. They still exist in the Stone Age or, in other words, in a very primitive civilisation.

This phenomenon is particularly relevant to Russia, which bridges the timocratical civilisations of the Western nations and the half-primitive civilisation of the peoples living in north-east Asia. This position obviously enlarges the outlook of Russia's diplomats and military people, forcing them to act according to known principles in completely different situations in the West and in the East.

On the one hand, the Russian military has to know how to wage war on its western border, which requires large armies equipped with all the technological means of contemporary civilisation. On the other hand, it also has to be capable of conducting extended campaigns by small units in the deserts and half-civilised states of Central Asia. While this significantly complicates the work of the Russian military, it also expands its experience and outlook, thus offering a fullness and comprehensiveness to its lessons.

The Aims of Contemporary Wars

Contemporary wars between civilised peoples pursue either (1) *an ideological goal*—for example, *national unification* (the Second Italian War of Independence in 1859, the Austro-Prussian War in 1866, or the First Balkan War in 1912) or *national unity*, as with the American Civil War in 1861–65; or (2) *an economic goal*—to conquer markets, capitals or trade routes, or wars in the colonies.

Sometimes, these clear goals can be characterised by a desire *to weaken the power of neighbours* (e.g., the Crimean War in 1853–56), in anticipation of a possible struggle in the future.

NIKOLAI MIKHNEVICH

A victory in war offers a right to life, opens opportunities for credit, and facilitates the civilising efforts of the victor. A conquest of land in half-civilised states opens an opportunity for colonisation, thus increasing the space for a potential increase of the population, as well as exploitation of natural resources.

The Connection between Politics and Strategy

The political motive of war defines the degree of its intensity, as the latter depends on the kind of *impression* that the former makes on the general population. Therefore, the characteristics of the people and their *historical instincts* have always to be taken into consideration.

Sometimes relations between two nations can be so tense, and hostile feelings so extensive, that a minor and insignificant political cause can spark a real explosion of hostility, leading well beyond the bounds of the initial insignificant cause.

On some occasions, *the political goal of war may correspond to the military one*. An example would be a war about conquering a defined territory. In other cases, the political goal itself may not constitute the object of war (as in the American Civil War in 1861–65 or the Austro-Prussian War of 1866). In these cases, it is important to define a separate military goal that can enable the victor to achieve his desired political aim.

If the general masses are indifferent to war, then the political object takes the lead, defining the extent of the mutual efforts of both politics and strategy.

Therefore, *a war can take any form, starting from the harmless deployment of an army of observation, and extending to wars of extermination.*

The ways of politics define the conditions of war. Politics sets the stage for war (*mise en scène*). It defines the strengths of the parties involved and the boundaries of the theatre of war.

Politics also defines relations with other states which are not involved in the war but are interested in its outcome. Their sympathy, or lack of it, can be of great significance in retarding or supporting the activities of war.

Politics plays a great part in defining *the moment for starting war activities*, as the way the war is waged depends a great deal on the right moment to start it.

Moreover, politics has a great work to do during the whole period of war, helping the military and using its successes to create a political situation unfavourable to the enemy.

After some successes on the battlefield, diplomacy can initiate *peace negotiations*. In this case, military activities should not stop under any circumstances, as decisive victory over the enemy has not yet been achieved.

If military actions on the battlefield turn out to be unsuccessful, the successful intervention of diplomacy can save the day. It can create an unexpectedly serious threat to the enemy or *find allies* that can tilt the balance of power against the enemy. In such situations politics gets priority over military operations, subordinating them. The Russians found themselves in such a situation by the end of the Russo-Turkish War in 1878, when they approached Constantinople.

Coalition Wars

If several states create a coalition, then their combined power will always be less than the sum of their powers. Without any doubt, all allies desire victory to achieve a defined goal. However, each of them attempts to transfer the most difficult job to the other. Moreover, in the majority of cases, each ally has his own thoughts regarding the desired outcome of war. While one may desire the complete destruction of the enemy, another may want to only temporarily weaken the enemy, forcing him to make concessions but preserving his power for the sake of the future.

Even after the establishment of a united command of the coalition forces, it is very difficult to separate the coalition's actions. Distrust, jealousy and intrigues are much more prevalent in coalition wars than in wars without allies.

Deploying the means of the allied states for the common interest is the main difficulty in coalition wars, as these means come under the control of different parties.

During a military operation, when *diplomacy takes an unwanted break*, military decision-making has to take the opinion of the allies into consideration. On some occasions, one will be forced to reject too bold a plan, so as not to scare the allies away. At other times, one will be forced to rush ahead with actions, to keep one's ally. Moreover, sometimes one will even be forced to reject military success, to avoid creating jealousy among the allies.

A general who succeeds in keeping his allies with him by a series of quick successes is lucky, because even *after few failures, the strength of the coalition will be in doubt*. The private interests of each ally will take the lead, sometimes even overriding the common goal defined at the beginning of the war.

In fighting against a coalition, it is important to find its weak side from the political as well as military point of view, and direct your blows against them. Under such circumstances, the political aspects frequently have significantly bigger potential than the military ones. Moreover, some ways of operating, usually considered in one-to-one warfare as enormous mistakes, will frequently be the wisest decisions in a war against a coalition. For example, when directing all your forces against one enemy, you can almost be inactive against another (like Prussia in 1757 during the Seven Years' War).

The Influence of Internal Politics

Internal politics also have an enormous influence on the process of war. *A weak government requires quick successes*, and therefore the space and time that it has to wage wars are more limited than those available for a strong government supported by an enthusiastic and energetic populace (such as Prussia in the Franco-Prussian War of 1870–71).

Moreover, in states *without a sole political authority*, military leadership is indecisive and the very preparations for war are easily misdirected.

To put it briefly, success in war *depends on the internal politics as much as it depends on the complete agreement between the foreign politics and military leadership*, which, in fact, also depends on the internal organisation of the state.

Knowledge of the weaknesses of the internal politics can direct the blows of a skilful and innovative enemy in the right places. Therefore, this explains why in times of danger, mature republics resorted to *dictatorship*, and why *a strong monarchy is the best form of governance in the interest of war*.

STRATEGIYA

The Economic Aspect of Contemporary Wars

The introduction of compulsory military service, the capitalist structure of states, and technological advancements have led to the emergence of the unprecedently large militaries of our time. Whole nations are getting ready to marshal their citizens under the banner to achieve their historical goals. Under these circumstances, conflicts can be resolved only by the complete exhaustion of the spiritual and material powers of each of the parties involved. [...]

II

The Plan of War

The Content of the Plan of War

The political situation and local conditions define the variables that constitute *the plan of war*, including all preliminary considerations regarding:

1) *the character of the forthcoming war;*
2) *the decision about the forces and means required to achieve the desired goal;*
3) *the creation of an opening situation from which military forces begin their actions* (strategic deployment).

The plan of war defines all the preliminary strategic operations (organisation of the army, preparation of the engineering aspects of the theatre of war, such as fortification and means of communication, mobilisation, transportation of forces and supplies, and, finally, the strategic deployment of the army). Implementation of all these in real life may take many years.

The Difference between the Plan of War, the Plan of Campaign and the Plan of Operation

All stated above parts of the plan of war should be prepared during peacetime and developed down to the smallest detail, as nothing has yet been affected by the enemy's intervention. As time passes and the situation changes, the plan of war should also be adapted and extended accordingly. The opening actions of a future commander-in-chief will be based on and limited by the initial concentration of military power, as defined by the plan of war.

For more detailed actions, military commanders will create their *plans of campaigns* and *operations*. While trying to follow the general guidance of the plan of war, they will ultimately be forced to *take into consideration the immediate situation*, i.e., the balance of power, the place, time, and the enemy's character. These variables can rarely be predicted in advance and, therefore, demand much creativity from commanders in the field.

This explains the famous phrase of Napoleon: 'I have never had a plan of operation!' Or the story about Suvorov in Vienna, just before his famous campaign in Italy in 1799, who, replying to a question about his plans, simply showed a blank sheet of paper.

A plan of operation depends on the enemy's actions, which are impossible to predict. Therefore, it is also impossible to draw up a plan of operation in advance.

During his operations, Napoleon used to march quickly to face his enemy. However, he repeatedly changed the exact combination of his forces, depending on the information he received, thus honing his deadly blow.

This clarifies the exact meaning of Napoleon's comment about his lack of an operational plan. *Such a plan can be drafted only in very general terms, outlining what we want to do and what we can hope to achieve with the means that we have at our disposal.* [...]

Defining the Main Idea of the Plan of War

As noted above, the main idea of the plan of war should include:

1) *the character of the forthcoming war;*
2) *the decision about the forces and means required to achieve the desired goal;*
3) *the creation of an opening situation from which military forces begin their actions* (strategic deployment).

1. The Character of the Forthcoming War.

A war can be either offensive or defensive, and in the latter case it can be either proactively or passively defensive. Either way, the character of war is defined, first and foremost, by the desired *political goals* and only after that by various *military considerations*.

While an offensive course of action offers significant advantages, it can be implemented only when the military completes its strategic deployment—in other words, when the military is prepared to act with the forces sufficient for the offensive. Otherwise, the idea of initial offensive action should be rejected, as France did at the beginning of the Franco-Prussian War in 1870–71.

An analysis and comparison of our and our enemy's mobilisation processes and concentration of forces will allow us, with an acceptable degree of certainty, to decide whether an initial offensive action is possible or whether temporary defensive actions will have to come first. [...]

2. The Forces and Means Required to Achieve the Desired Goal.

Superiority over the strength of an adversary is the guarantee of victory. However, it is important to remember that this superiority is defined not only by the physical size of the military forces, but also by the spiritual (moral) component. Napoleon, defining the importance of the spiritual component in war, stated: *'In any military enterprise, three-quarters of success depends on the moral factor, and only one quarter depends on physical strength.'*

Moreover, victory also depends on *the skilful deployment of forces*. There is no need to seek superiority always and everywhere. It is enough to achieve the desired superiority in a *decisive place* in *a decisive moment*, securing a *local victory* that will definitely have an influence on the whole operation. [...]

A clear understanding of the situation, boldness in decision-making, skilful calculation of time and space, economy of force (i.e., a skilful concentration of forces according to the importance of the terrain), etc.—all these are the means that allowed the great masters of military affairs to suddenly achieve the required superiority of forces in a certain place and at a certain time, thus securing victory over enemies who were more powerful but less energetic, less active or less capable.

The strength of the military is not defined only by its size. Instead, it is an outcome of many complex variables of a physical and spiritual character. It seems right to argue that the most important of them are:

NIKOLAI MIKHNEVICH

A. Size and organisation of the military
B. Moral characteristics of the soldiers and their commanders
C. The resourcefulness of the high command
D. Training
E. Weaponry.

A. Size and organisation of the military.

Attempting to calculate *the size of the enemy's military* should not be a problem. After all, under the parliamentary system of European countries, all matters of recruitment and military organisation, as well as statistical data regarding population and industry, cannot be kept secret.

The military should be organised in a way that will turn it into a convenient tool in the hands of commanders. Military units should not consist of more than four or five subunits. Napoleon used to say that 'dealing with four or five subordinate commanders is the limit of human capability in battle'. Similar divisions into four can be seen from the time of the Macedonian phalanx, through the battalions of Gustavus Adolphus, and up to the contemporary organisation of companies, battalions and brigades. History shows that any violation of this basic organisational principle will have serious consequences. [...]

B. Moral characteristics of the soldiers and their commanders.

The question of the moral characteristics of the soldiers and their commanders is much more complicated than the question of size or organisation. The best way to approach it is through the idea of an *energy coefficient*, or *the numerical expression of the spiritual resilience* of the military in difficult moments.

A close examination of military history shows that young nations, which have just entered the stage of history, have the highest degree of military energy and valour. However, as time passes and these nations become infirm, all their vital functions get weaker, including their capacity to stare calmly into the eyes of death. Therefore, we can see throughout history that the older nations become, the fewer casualties they can suffer in battle without losing their capacity to continue fighting. [...]

STRATEGIYA

Without any doubt, the more civilised nations are, the less militancy they contain, which ultimately reduces their spiritual resilience during war. This fact is supported by history—for example, the ancient peoples who fell under the blows of a younger civilisation (Persia), or Rome, which was conquered by significantly younger peoples, the ancestors of the present-day nations of Western Europe. Russian people joined the stage of history almost a thousand years later and, therefore, they are younger in their progress in civilisation, they have stronger nerves and a greater capacity to overcome the hardships of war. For example, Russian forces were able to continue fighting regardless of casualties of thirty to forty per cent, and regardless of the personality of their commanders, whether it was the beloved Alexander Suvorov and Mikhail Skobelev,[2] or the much-hated William Fermor.[3]

Therefore, this variable should be considered as a constant, which is defined by the inherent characteristics of the contemporary Russian population. In their wars against Russia, European armies had to invest much more effort in comparison with the wars they fought among themselves. This is because they had to overcome the significantly stronger energy and willpower of the Russian people. [...]

Consequently, it seems right to assume that *every military has its own level of endurance, and, therefore, in comparing the size of militaries, it is important to apply a certain coefficient, which can be defined by using statistical data from recent wars*. These coefficients are of great importance in war and should be taken into consideration at the strategic as well as tactical level. [...]

The moral characters of commanders should be even higher than those of their subordinates. This is because in addition to physical stress (to which all soldiers are exposed), they are also required to have enormous composure. This is why the Romans used to say that 'a herd of sheep headed by a lion is better than a herd of lions headed by a sheep'.

---

[2] Mikhail Skobelev was a Russian Imperial general famous for his heroism during the 1877–78 Russo-Turkish War.
[3] William Fermor was a Russian Imperial general, best known for his command at the 1758 Battle of Zorndorf during the Seven Years' War.

In modern-day paramilitary armies (with very short periods of compulsory service), the quality of the commanding personnel is even more important than in the past. [...] Large contemporary armies require commanders with the best characters and staff officers of the highest quality, because the reality of war will require them to make decisions without clear orders and take responsibility in the context of general goals defined by their superiors.

C. The resourcefulness of the high command.

Nowadays, the resourcefulness of the high command and the preparedness of military staff officers at headquarters are especially important. In addition to talent, the contemporary deployment of enormously large forces requires wide technical knowledge and a deep scientific background. As ever larger forces are deployed and as the conditions of time and space become more complex, it becomes more difficult to undo any mistakes.

The talent of the general has the largest influence on the conduct of war. '*People in war mean nothing as everything is done by one person,*' argued Napoleon. 'The first quality of a general', he used to say, 'is his cool head (avoir une tête froide) capable of forming the right impression and incapable of being irritated or muddled, or overwhelmed by good or bad news; it has to be able to assess all consecutive or simultaneous events, giving them the exact attention they deserve, as *glazomer*[4] (le bon sens, la raison) is about comparing and judging many impressions under similar conditions.'

In other words, a general should have a *balanced mind and character* (l'homme carré, according to Napoleon). With a bright and clear mind, he should be distinguished by an ability to take the boldest decisions. [...] According to Clausewitz, '*boldness governed by superior intellect is the mark of a hero*'. [...]

D. Training.

The outcome of *coherent training* and education of forces during peacetime will obviously manifest themselves during war. However,

---

[4] The Russian word *glazomer* (*coup d'œil*—literally, an ability to measure distances by eye) describes the ability to assess a situation.

while in war units will do what they were trained for, they will perform significantly worse than in peacetime, because: (1) the conditions of battle will unsettle soldiers' minds; and (2) half of the personnel, especially in the infantry and artillery, will be reservists, who will be neither properly trained nor familiar to their commanders.

E. Weaponry.

Weaponry should also be taken into consideration, as the enemy's defeat and its fighting spirit are determined, to a certain degree, by the weaponry used against it.

It is important to do everything possible to avoid any inferiority of weaponry relative to the enemy. One of the most significant guarantees of victory—forces' fighting spirit and self-confidence—depends on that. Differences in weaponry can influence the forces, and the masters of military affairs have always paid special attention to them.

To conclude the discussion of the different elements that define the strength of the military, it seems right to argue that this question is very complex. As stated above, it is wrong to assess only the physical size of the opponent's military, as the moral characteristics of the soldiers and their commanders, the resourcefulness of the high command, training and the quality of weaponry—all these should be taken into consideration as well.

Any difference between two opposing armies in one or other element of their strength should be taken into consideration by applying a certain coefficient, defined as the combination of differences between these elements.

Without any doubt, this is not an easy task. However, only people who do not understand the essence of military affairs assume that solving military questions is an easy task. Military geniuses who have obvious answers to these questions are as rare as geniuses in any other field of human activity.

3. Creation of an opening situation from which military forces begin their actions (strategic deployment).

The last and most important element of the plan of war concerns the question of the concentration of force before the beginning of war, the so-called *strategic deployment of the military*.

NIKOLAI MIKHNEVICH

Strategic deployment has to do with the initial position taken for the operation. This is such an important decision in war that it can frequently shape not only the whole progress of war, but also its outcome.

Given the large theatres of war and enormous armies of the present day, any mistakes made during the initial strategic deployment, according to Moltke, 'can be fixed only with difficulty during the campaign'.

First and foremost, the strategic deployment should be planned and conducted in a *secure* manner. In other words, it should be carried out without any need to bend to the enemy's will, creating, at least in the main theatre of war as anticipated, a favourable situation for conducting operations (i.e., *favourable conditions in terms of the balance of forces, place and time*). [...]

Developing the Plan of War during Peacetime

All details of the plan of war have to be developed in peacetime. This development should be based on various assessments in terms of *the goal and direction of military actions*, i.e., according to *the operational line*, as it used to be called.

The question of the operation's goal and direction (the choice of operational line) is one of the most important questions of strategic art.

In the past, when preparations for war did not require significant long-term activity, the question of the operation's goal and direction was frequently elaborated by one person, who would also be responsible for its execution. Nowadays, however, such a favourable situation is a true rarity. Contemporary plans of war and choices regarding operational lines are sketched out long before war begins. Moreover, it frequently happens that the commander-in-chief will be forced to start his operations in a pre-existing situation, with the aim of achieving goals that were defined for him long ago. Therefore, to avoid unforgivable mistakes, all aspects of the operation's goal and direction have to be carefully thought through in advance.

Contemporary wars are fraught with dangerous consequences. Therefore, contemporary militaries (i.e., armed nations) and contemporary strategic art should involve a true collaboration of the best intellectual capabilities of the military. Everyone involved in the development of the plan of war should be responsible for any major mistake in

STRATEGIYA

the conduct of war. This is why the main principles of strategy should be accessible to as large a circle of military intelligentsia as possible.

It is important, therefore, to focus on the question of the operational line. *The choice of operational line* consists of two factors:

1) The goal of actions (the objective of the action).
2) The direction of actions.

1. The Goal of Actions.

The goal of a campaign *must be important*, otherwise, according to the creator of contemporary strategic theory, Henry Lloyd, '*even ten campaigns will be useless*'. There are three possible goals of action:

A. Destruction of and victory over the enemy's military.
B. Conquest of certain territory.
C. Certain political results achieved by military means.

A. *Destruction of and victory over the enemy's military* will most often compel the enemy to accept our will. However, we should remember that contemporary civilised nations have access to very powerful resources and can recover their strength in a relatively short period. France during the 1870–71 Franco-Prussian War provides a good example of this capability. After losing its army at the beginning of the war, it took France no more than six weeks to deploy approximately 600,000 troops against Prussia, i.e., almost twice the size of the original army deployed at the beginning of the war.

Therefore, it would be wrong to assume that victory over the enemy's military, or even its destruction, will ultimately lead to a successful conclusion of a war. It will also require:

B. *Conquest of important centres of the enemy's civilian and governmental activity*, i.e., conquest of the enemy's capital and other places with the greatest concentrations of population. Only by conquering the enemy's state is it possible to prevent the enemy from continuing the war.

In countries that are not advanced in civilisation, the conquest of the capital can be sufficient, as in our wars in Central Asia. However, in countries with large territories and well-developed local government, the fall of the capital will not necessarily be followed by the fall of the government, thus allowing the enemy to continue its struggle. For example, Napoleon's conquest of Vienna in 1805 and 1809 did not bring peace; it was his victories at Austerlitz (1805) and Wagram (1809) that forced the Austrians to the negotiating table. Moreover, his conquest of Moscow in 1812 did not secure his victory, and his inability to conquer the country led to the ultimate failure of his victorious campaign in Russia.

Consequently, subjecting a contemporary civilised state to our will requires us: (1) to defeat its military forces; and (2) to conquer the vital centres of its civilian and governmental activity.

C. Sometimes, the situation can be so managed that there is no need to pursue such a large goal as the enemy's complete destruction. Such a situation frequently characterises coalition wars, when *a political goal* is sufficient; for example, holding down the enemy forces and distracting them from the actions conducted by one's allies, thus making their task easier.

The best example of the influence of *such political considerations* on the definition of the operation's goal was the Italian theatre of war during the War of the Polish Succession. In 1733, the commander of the French forces, Claude Louis Hector de Villars, did not seek to expel the Austrian forces from Italy. Instead, he sought to keep them in Italy, thus preventing them from withdrawing to reinforce the main Austrian force on the Rhine. He realised that the French allies, the Spanish and the Sardinians, would not follow the French as far as this, and defeating the Austrians (allowing them to withdraw) would only reinforce the troops that the main French Army had to face alone.

2. The Direction of Actions.

The direction of actions should be:

A) *the shortest*, in relation to the objective of the actions {While the enemy's army is the most important objective of military actions, predicting how it will deploy its main force is not always possible. Therefore, the operational line should be directed towards other goals, such as the enemy's capital or other vital centres of its civilian and governmental activity, assuming that the enemy must protect them with its military.};

B) *the most convenient* from the operational and administrative point of view; and
C) as *secure* as possible.

A. Shortness.

The shortest way (the shortest operational line) leads to the fastest achievement of one's goal, offering the advantage of time, which, under favourable conditions, provides one with initiative, which is very important in war. The shortest operational line shortens *the supply lines*. Therefore, making the operational line as short as possible is very important. For example, Lloyd argues that the length of the operational line is the first matter to be considered in planning. {He argued that the operational line should be: 1) shortest; 2) most convenient; and 3) directed to the goal.}

B. Convenience.

The chosen direction of action should be *the most convenient from an operational and administrative perspective*.

From the *operational* (tactical) perspective, the direction should assist in removing *as many obstacles as possible from the way of all the main or most important forces*. Moreover, the chosen direction should not complicate communication lines, freedom of movement, and the fighting activities of the forces deployed. […]

From the *administrative* perspective, the chosen direction should lead into territory with developed *railways* and *water transport*, without which contemporary large armies can neither exist nor operate. {Only small deployments are possible without railways.} The direction of action that leads through highly populated areas is especially advantageous, as they offer easy quartering for troops, allowing them to use various local facilities.

C. Security.

Security of the direction of action is vital, and it depends on local conditions, political circumstances, the length of the operational line, and the quality of manpower.

The security of the direction of action requires one to conduct operations under as favourable circumstances as possible, thus trying to reduce risks as much as possible. This includes all means and actions that lead towards the concentration of the required force in the decisive place at the decisive time; and the protection of the army's rear, i.e., its communication and supply lines. [...]

III

The Execution of the Plan of War (in contemporary conditions)

Organisation of the Military

'Organisation predetermines life and activity'

Aleksander Stronin

The quality of forces is measured by their ability to fight and manoeuvre—in other words, by their ability to move fast and easily with all the necessary supplies and provisions. Obviously, the quality of forces depends a great deal on their natural characteristics (physical and spiritual). However, it also depends on a skilful combination of all elements of military strength. This combination is, in fact, the main goal of military organisation and training. *Good organisation of forces* has an enormous importance, as it allows for the reduction of friction between different units and personalities, thus helping to develop more productive actions. *The principles of organisation are*:

1) Unity of command.
2) Convenient command.
3) Independence of force.
4) Close internal relations between different parts.

1. Unity of Command.

'The unity of command is a requirement of the highest importance.'

Napoleon

'One mediocre commander is better than two good ones.'

Henri de La Tour d'Auvergne, Viscount of Turenne

Accepting unity of command, as a mean to achieve unified action in war, is the most fundamental principle of military organisation. However, there is a limitation on the number of people that one commander can supervise. *It is necessary, therefore, to recognise the power of secondary commanders*, who have to carry out the will of the senior commander and ensure its execution in the context of various eventualities, which may occur over large spaces occupied by separate elements. This leads to the necessity of grouping forces into large and small units, as well as the training of high-level *secondary commanding personnel*.

The skilful command of forces has frequently proved itself one of the decisive factors in victory. However, regardless of how skilful and talented the commander is, he needs a similarly skilful contribution from his secondary commanders, i.e., his assistants in the difficult task of command. [...]

Therefore, military affairs, like other complex affairs, require the application of *the principle of the division of labour*: the decisions of the senior commander should govern all his assistants, all commanding elements of the military. This leads to the obvious conclusion that the commanding element of the military *manifests its intellectual and moral strengths*. The closer a commander is to the senior commander, the more intelligence is required from him. The closer he is to the soldiers, the more decisive he should be.

It is in the interest of the high command that the whole chain of command be trained in the spirit of initiative. In other words, all commanders should be able to act independently, without direct orders from their superiors in pursuit of the common goal, especially when the situation requires this and there is no time to receive orders. {Peter the Great had always offered much room for initiative to his subordinates, ordering them to act 'as is required of an honest officer and good person'. He forbade them to address him with minor questions, demanding that they 'decide according to local requirements, as anticipating orders does not stand the test of reality'.}

While Clausewitz considers the simple execution of an order as a sign of mediocrity, Colmar Freiherr von der Goltz[5] calls such behaviour an insufficient understanding of an officer's duty, arguing that *an officer who is guilty of inaction cannot justify himself by claiming a lack of orders.*

---

[5] Wilhelm Leopold Colmar Freiherr von der Goltz (1843–1916) was a Prussian field marshal and military writer.

NIKOLAI MIKHNEVICH

In addition to the required character, initiative also requires a proper understanding of military affairs. While a wrong initiative can sometimes contradict the common goal, it happens significantly less frequently than the damage caused by the lack of initiative. Moreover, as Goltz claims, initiative has very powerful enemies to struggle with, such as spiritual lethargy, apathy, routine, fear of responsibility, as well as the general tendency of people to be governed by unfolding events and wait until they are shown what to do, instead of acting according to an understanding of their duty. These negative qualities easily paralyse any desire for initiative.

There are positive ways that allow one to mitigate any possible detrimental aspects of initiative. One of the most important of them is *intellectual discipline within the army*, which requires the existence of *a solid military school*. This, however, should be understood in its wider context. It does not mean that the actions of commanders should be systematised according to a certain set of rules. Instead, it means that *a certain common understanding should govern ways of solving problems, when certain common principles are instilled in the blood of military commanders*. [...] When the army has intellectual discipline, then a similar problem, given to different officers, will have different solutions; yet, regardless of how different they are in their details, they all share the same essence.

On the one hand, the large theatres of war and massive armies of the present day require the complete coordination of activities in all directions within the general plan of the senior commander. On the other hand, the same large distances require a similarly *great degree of independence of the commanders*, otherwise there will be no time to execute all orders of the senior command.

Any army should be headed by a *senior commander*—a person who has the ultimate power. As Suvorov used to say: '*Full power to the entrusted commander.*' The nature of war requires that the senior commander will enjoy full command of his powers and be trusted not only by his forces, but also by the general population. [...]

In principle, *the senior commander should only take decisions whereas their development and execution should be performed by his subordinates. The chief of staff* should be the person responsible for translating the will of the senior commander, and therefore he should be selected by the senior commander from among a number of people with whom he has feelings of friendship and trust. Moreover, *for the tasks of a personal or secret character*, the senior commander should be served by several officers with undoubted loyalty and diligence. [...]

2. Convenient Command.

For the convenient command of the enormous armed forces of contemporary first-class countries, they can be divided: (A) according to the type of the forces; (B) into military tactical organisms (armies, individual corps, etc.) and organs (corps, individual divisions, etc.). The proper combination of these, according to the requirements of any given situation, allows for the utilisation of the whole potential of the army.

A. The Division of the Military according to Types.

Contemporary armed forces are created from several types of forces, each of which has its own special purpose, determined by the requirements of contemporary war. Almost every contemporary army includes four main types of forces:

i) *Active Service Forces*—intended to act on the battlefield in pursuit of major strategic goals.
ii) *Reserve Service Forces*—intended to reinforce the active service forces, as well as perform duties in the rear.
iii) *Reserve or Reinforcement Forces*—intended to reinforce the active or reserve service forces or to form new units.
iv) *Administrative Forces* (of a supportive nature)—they do not have any battle purpose, but they have great significance in peacetime and in war (supply, medical, railways, logistical and other units) [...]

In addition to these four main groups of forces, there are two other, which are used in the rear or for various special activities:

i) *Militia*—less prepared than forces deployed on the battlefield, but can be very useful in defensive or fortification war.
ii) *Partisan Forces* (civilian uprising)—armed parties that act as individual units, either with or without direction from the main forces.

Even though this division of armed forces by type significantly simplifies their deployment in war, in contemporary wars any senior commander can find himself commanding more than a million troops. Therefore, to simplify his duty even more, it is necessary to divide the forces into units, thus allowing their most beneficial deployment in battle.

B. The Organisation of the Military.

The armies of the future will consist of millions of soldiers, and will probably be divided into several *individual armies*. Moreover, with further development, the armies of the first-rank states will probably consist of several *groups of individual armies*.

The experience of previous wars suggests that an individual army should consist of between 150,000 and 200,000 soldiers, divided into four or five corps. Having more than five corps is not recommended, as, to quote the words of Napoleon, 'owing to the limits of human ability, no general can command more than five units in one battlespace'. After 1800, he always followed this rule. [...]

3. Independence.

The independence of a military unit is defined by the extent of its ability to act in the theatre of war without support from other military units and institutions. Therefore, the independence of a military unit is defined by its tactical and administrative independence.

The tactical independence of a military unit is defined by *its size* and *the inclusion of different types of forces*.

Administrative independence depends on the *size and type of the attached battle-support units (transportation, medical, supplies, etc.)*.

While it seems desirable to make even small units independent, this would create a fragmentation of administrative institutions, leading to an undesirable increase in the administrative element within the army. On the other hand, a preference very large independent units would lead to an excessive centralisation of battle-support services. It would require the movement and deployment of large battle-support units in close proximity to the troops, slowing the deployment process and offering no assurance that the troops would have everything they need in the right place and time.

These considerations lead us to conclude that in a large-scale war the battle-support units should be attached to units the size of a division or corps. The choice of all the main European armies has fallen on the corps.

However, this does not do away with the need in other types of war, such as mountain wars, *for divisions or even smaller military units to be supplied, temporarily or permanently, with everything they require to manoeuvre independently*. It is also possible to attach some battle-support services to units that are larger than corps, e.g. an army. While various services, such as railway building and maintenance or engineering units for besieging large fortified points, are required to ensure an army's independence in the theatre of war, no corps has any real need for these types of units.

In conclusion, within contemporary large armies, army commanders mainly deal with and give direct orders to corps. Consisting of four different types of forces, *a corps* has significant fighting power and is provided with all the means required for independent deployment. Following this organisational principle, a corps constitutes a unified organ in which each element has a very specific purpose within the whole, while all elements are harmoniously subordinated to a single command. Therefore, it is possible to argue that a corps is *a strategic unit of contemporary warfare*.

On the one hand, individual cavalry divisions should also be able to accomplish missions independently. On the other hand, they can take with them only absolutely necessary elements, as their speed and manoeuvrability determine their success. Cavalry divisions, burdened with large supply units, would be of little help to the army, probably causing more harm than benefit. On the contrary, their organisation should allow them to operate on the front line, living as much as possible on the supplies available in the theatre of war. [...]

Corps, as elements of an army, should be deployed relatively close to the army commander, so they can be commanded *by definite orders given every single day*. Moreover, in difficult circumstances they can be supported by reserve forces or the personal involvement of the army commander.

Individual armies, however, operate under completely different conditions. The general command cannot command individual armies by means of definite orders. Instead, the general command should focus the attention of the commanders of the armies on the general idea of the operation and the general situation in the theatre of war.

Therefore, in the majority of cases, the commanders of individual armies will be required to issue daily orders by themselves. This will obviously require a very strong proficiency in independent decision-making. In other words, it will require *a highly developed sense of initiative, an ability to grasp the strategic situation, and self-orientation* in the whole theatre of war. An army commander will be required to have *a conscious strategic sense*, which allows him to comprehend the strategic operation in its entirety [...] and distinguish between its very important, less important and unimportant aspects, concentrating all his efforts exclusively on the very important, and abandoning the rest. [...] While a corps commander is good if he, according to Napoleon, 'can wage a war on large roads and under artillery fire', an army commander should also 'be able to wage it on the map'. As Suvorov put it, he 'should be able to see the war as a whole', i.e., be a strategist.

4. Preserving organic relations within the forces.

Close internal relations between subunits of one military unit ensure strong *discipline, mutual trust* and *responsibility to the unit*, creating a high degree of *cohesion*, which lies at the heart of military success.

The lack of cohesion in *militia (improvised) forces*, combined with their unfamiliarity with the hard work and deprivations of war, makes them unsuitable for prolonged offensive operations, though they are quite suited to defence.

These different characteristics of permanent and militia armies define the different ways they can be used by both strategy and tactics.

Preserving the organic relations within forces has paramount importance in times of war. While particular situations will require different types of deployment, it is important to avoid dissolving organic units to create assembled units (brigades or divisions), which will never be able to perform at the level of a well-established organic unit, where every person knows the other, and the commanders can allocate orders according to the capabilities of their subunits. [...]

IV

Executing Military Actions (Operations)

Major and Secondary Operations and the Relations between Them

Every large war consists of several *military operations*. Some of these are *major operations* that have a decisive influence on the whole direction of war. Others—*secondary operations*—have no decisive influence on the general direction of war. Their role is to provide the required support to the major operations, securing their success. [...]

Assembly of Force: The Most Important Task of the Strategist

The art of the strategist should express itself in his ability to *assemble a force* in such a way that the part assigned to major operations is as large as possible, while the part assigned to secondary operations is as small as required. Making this decision is a very difficult task, and even the greatest masters of military affairs have occasionally failed to address it correctly. [...]

The Main Conditions Which a Plan of Operations Has to Fulfil

Any plan of campaign or operation, rooted in the main goal (either defensive or offensive) pursued in a given theatre of war, should meet the following conditions:

1) Lead to the achievement of an important goal;
2) Be convenient;
3) Be safe, from the perspective of cooperation between the front line and rear forces {This condition of safety implies: (1) facilitating as complete a concentration of force as possible; (2) economy of force—the skilful deployment of forces for secondary aims; (3) facilitating the coordination of actions; and (4) rational implementation of personal initiative.};
4) Meet the conditions of the given situation (force, place, time, the enemy's will);
5) All episodes of the operation (battles, sieges, deployment of forces to given locations) should correspond completely to its main idea (unity of operation—its internal integrity).

Unity of Operation: Its Internal Integrity

Unity of operation, or its internal integrity, is expressed in the ability to pursue one main goal. *It is the distinguishing characteristic of the strategic art of the great masters of military affairs.* [...]

NIKOLAI MIKHNEVICH

Plan of Operation Should Correspond to the Character of Its Implementer

It is obvious that in pursuit of its main idea (the goal of a given campaign), *a plan of operation should correspond to the character of its implementer*. As Napoleon put it: 'In military art, like literature, everyone has his own preferred genre.' [...]

The Gift of Foresight and Luck

The gift of foresight—*the ability to sense what will take place*—which marks the great masters of all arts can be inherited. However, as with any other human ability, it can be developed by relevant training and exercise.

Luck and fortune, obviously, can help a lot. However, just as any accidents in war obey certain known rules, luck and fortune also do so. *An understanding of their affairs allows the chosen ones 'to rule their luck'*. [...]

Waging Operations: Main Conditions

The following conditions are required to achieve, as much as possible, the aims of any operation:

1) To keep affairs secret as long as possible.
2) To know as much about the enemy as possible by maintaining constant contact with him.
3) To secure as complete a concentration of force as possible.
4) To dispatch units for secondary missions only when absolutely required.
5) To maintain coordination between the actions of different forces.
6) To act decisively.
7) To act fast, but not without judgement.
8) To act persistently.
9) To act with maximum energy.
10) To be vigilant.

1. Secrecy.

Surprise is the best way to achieve success in war, but it is a consequence of keeping all related preparations in secrecy. For example, during military operations, Napoleon prohibited private newspapers from publishing anything relating to the army or navy. They were only allowed to republish articles from the official 'Le Moniteur'.

While this secrecy is relevant in general terms, it has no place when the troops start to perform their tasks. They must know what is required of them. As Suvorov put it: 'Every soldier has to understand his manoeuvre.' Only in such conditions can everyone comprehend his goal, and success will accordingly not depend on the direction of a random bullet.

2. The necessity to maintain constant contact with the enemy.

Intelligence about the enemy is one of the main sources that help one to understand the situation in war. It can be gathered and updated only by maintaining constant contact with the enemy. Usually, this will be the task of the cavalry, which will be sent behind the enemy lines to try to penetrate the defensive curtain of the enemy's cavalry.

Accidental battles and many other catastrophes in war are the consequence of a failure to maintain constant contact with the enemy. [...]

3. Concentration of force and the choice of the right place.

Concentration of force, as complete as possible, in the decisive place at the decisive moment of operation has the best chance of securing success. However, the most important aspect *is the skill with which this choice is made*. [...]

4. Forces for secondary missions.

Dispatching forces for secondary missions should be allowed only when absolutely necessary and their size should be kept to an absolute minimum. It is important to remember that the basic rule of military art is the principle of the concentration of force. [...]

5. Coordination among forces; personal initiative.

Coordination among forces—unity of action in achieving a single common goal—is the first rule of success in war. Only good coordination can allow the full exploitation of the means and forces required to achieve victory over the enemy.

The question of coordination goes hand in hand with the question of *personal initiative*, which is one of the most important and proactive elements in military affairs. One might assume that good coordination of forces manifests itself in precise and unconditional subordination to the will of the senior commander, thus contradicting the very idea of personal initiative. In practice, however, this is not the case, as both are equally important.

While the commander's orders should always be precisely fulfilled, *they should be fulfilled according to their spirit, and not their letter*. Suvorov made several statements to clarify this issue:

'Owing to his proximity to the situation, a local commander is in the best position to judge; he comprehends hourly changes in the situation, and therefore can direct his actions according to military principles.'

'If I say the left, but it should be the right—do not listen to me.'

'If I say move forward, but you see better—do not go forward.'

It is important to remember that personal initiative should mostly be used in the field of tactics, where a favourable situation can exist for only few moments.

In the field of strategy, however, the question of timing is more flexible. Therefore, especially with modern means of communication, it is possible to exchange opinions, and ask permission to perform an action that seems to contradict the received order.

Individual armies and corps, independently deployed in the theatre of war, can receive significant freedom of initiative. However, formations that act in close cooperation should seek to follow orders as precisely as possible.

Any initiative of the commanders should be based, first and foremost, on their understanding of the situation, commonality in their ways of addressing challenges on the battlefield, and close coordination among their forces.

While this coordination is primarily based on various physical connections between forces, it should, first and foremost, be rooted in the shared values of a conscientious approach to duty, solidarity and mutual help.

Coordination of actions and the proper direction of personal initiative among subordinates (as well as their training to execute this initiative) should be one of the major duties of any commander. He should meet with them as frequently as possible, instruct them as to the desired goal as accurately as possible, and inform them about the overall situation as correctly as possible.

Subordinate commanders should also do their best, maintaining contact with their senior commander, as well as with their neighbours on the battlefield, sharing their intelligence and other information regarding their situation. Without this difficult and meticulous work, no coordination among forces is possible.

6. Decisiveness of action.

Decisiveness—the ability to take bold decisions—is one of the qualities required of any military commander. Taking bold decisions is not an easy task, as the full range of circumstances is not always known. Moreover, the decision can be obstructed by various delaying factors, such as the fear to take responsibility or the need to carry out the original order. However, a true military officer should probably fear more being accused of indecisiveness than taking bold decisions. [...]

7. Swiftness of action, but not without judgement.

Swiftness and impetuosity of action significantly increase the chances of success, as they prevent the enemy from conducting the necessary countermeasures. As the famous aphorism states: 'Every manoeuvre will be answered by a counter-manoeuvre, so the right minute should not be missed.' While this aphorism underlines the importance of speed in conducting operations, speed of action *should be distinguished from thoughtless haste.*

When we speak of the swiftness of action during operations, we mean these actions should be the result of: (1) swiftness in thinking while drawing up the plan of action; (2) swiftness in giving orders and conducting preparations; (3) swiftness in deployment of forces; (4) speedy exploitation of the right moment and all other variables in a given situation; (5) not contradicting orders, etc.

8. Persistence.

Persistence in pursuing a goal is extremely important for achieving success under the difficult circumstances of war. In addition to the personal abilities of the commander, persistence in war requires corresponding endurance on the part of the forces and their appreciation of their commander. Only a combination of all these conditions could create the famous examples of persistence in war, such the 1796 Battle of Arcole, the 1799 Battle of Trebbia and the 1812 Battle of Borodino. [...]

Usually, when lacking battle experience, commanders start to consider their defeat too soon. While appreciating the difficult situation of their own forces, they frequently forget that the enemy's situation might be even worse. However, military history shows that, unless all forces surrender to panic and run away from the battlefield, victory still can be achieved. The question is which side can endure for longer, which side can avoid losing its head and achieve the required level of persistence and an unlimited desire to win success at any cost.

9. Energy.

War strains the physical and spiritual powers of military forces and their commanders, as war is the touchstone of their energy as well as of the energy of the nation that created them.

Napoleon used to say proudly about his forces: 'March for forty kilometres a day, fight and only then rest—I do not know another way to wage a war.'

Such energy can only be a result of cooperation between the commander and his forces. They must deserve each other. In the hands of one commander, units can perform differently from the way they perform in the hands of another. [...]

Forces can work energetically only when they understand their commander, trust him, and see in him an expression of their spirit, as well as of their national ideals. When a commander expresses the spirit of his forces, they will follow him into any dangerous affair, and will not stop in the face of any sacrifice or effort.

To unleash this kind of energy, a commander should distinguish himself by significant energy and health. His spiritual energy, however, is much more important than his physical health—something that was well proven by Julius Caesar, Napoleon and Suvorov. While they were not gifted with great physical health, their willpower had no limits. They were people of a demonic nature, capable of living by their willpower alone. [...]

10. Vigilance.

Vigilance or *caution*, in the best understanding of this word (i.e., caution that comes not from fear, but from a desire to counteract unfavourable incidents), is required from military commanders of all levels.

STRATEGIYA

It should manifest itself in both peacetime and times of war (especially during battle). It expresses itself in the *preservation of all means required to maintain military activities* (food, ammunition, medical supplies, etc.), or in a *resolute execution of all the requirements of military life* (maintenance of discipline, scrupulous attention to communication, deployment of guards, and intelligence gathering).

While all these requirements seem to be obvious, and no military commander would disregard such important issues, the reality often proves to be different. In practice, these requirements are expressed in an enormous number of small duties, required to be repeated every day and thus liable to become tedious. On the one hand, the importance of these requirements increases in times of war. On the other hand, physical and sometimes spiritual fatigue, so prevalent in times of military operations, frequently weakens the vigilance needed to perform these duties.

In contemporary conditions, when wars are rare and more than a half of all military forces consist of reserve units which are not accustomed to military life, military commanders should demonstrate an extreme vigilance in regard to the actions of their units. [...]

4

THE PHILOSOPHY OF WAR[1]

Anton Antonovich Kersnovski

PART ONE: ON THE NATURE OF WAR

[...].

Chapter 2: The Definition of 'Justice' and the Aims of War

When the state and nation are deified, the only criterion for judging the degree of justness of a given war is the degree of its benefit for this state and nation. If someone who decides to take up the sword and considers war the only way to defend his legal rights, then there is no way to make him doubt the validity of his claims. The 'Manifesto of the Ninety-Three'[2] is, in this respect, one of the documents that characterises humanity best.

---

[1] This chapter is based on the translation of selected chapters from Anton Kersnovski, *Filosofiya voyny* [The Philosophy of war], (Belgrade: Izdanie 'Tzarskogo Vestnika', 1939).

[2] The 'Manifesto of the Ninety-Three' is the name commonly given to the 4 October 1914 statement endorsed by 93 prominent German scientists, scholars and artists, declaring their unequivocal support of German military actions in the early period of the First World War.

STRATEGIYA

The first type of wars—those that are waged in defence of the highest spiritual values—are absolutely just. All our wars in defence of fellow Christians against Turkey or Poland, as well as the 1917–22 Civil War, from the White side, belong to this category.

The second and most common type of wars includes wars waged in the name of the state and nation. There is no general rule that can measure the justness of these wars. Every case requires its own evaluation, and in every case the assessment will be subjective.

The third type includes wars that answer neither the interests of the state and nation, nor the requirements of higher justice. These wars can generally be categorised as selfless or even meaningless adventures, such as Russia's participation in the allied wars in 1799 and in 1807–17, or its involvement in Hungary in 1849, or the French intervention in Mexico led by Napoleon III.

Wars of the first (absolutely just) and third (absolutely unjust) types are relatively rare. Most gunpowder was burned and blood spilled in wars of the second type—wars driven by statehood and nationalism.

As noted, there is no general rule for assessing the second type of wars. Therefore, before analysing each single war, we need to collate them, grouping together all wars between two given nations and tracing their relations throughout the centuries. In this way, moving backwards through history, we will be able, sooner or later, to arrive at the original cause (the root) of their confrontation. Only then will we be able to define who 'took up the sword' first, thereby disturbing the original harmony between these two given states and nations.

To do so, we need to reject any sense of chauvinism—a feeling that anyone who loves his motherland should avoid as much as possible if he does not want to call disasters upon it. Keeping this in mind, let us analyse the justness of the Russo-Polish wars in the general historical context of Russo-Polish relations.

In the beginning, the relations between these two Slavic peoples were neighbourly. The initial reason for conflict occurred in the 13th century, when the Polish kings laid their hands on Red Ruthenia and then on White Ruthenia, creating a disenfranchised 'rayah' of the Russian Christian Orthodox community inside the Polish state—much-suffering future 'dissidents'. Therefore, the centuries-long Russo-Polish confrontation began at the hands of the Polish. The Union of Lublin and the adventure of Sigismund III of Poland which resulted in a temporary occupation of Moscow—all these were the subsequent stages of Poland's offensive and oppressive movements against Russia.

As the Russian state became stronger, the Polish one was declining. The subsequent First Partition of Poland of 1772, which, in fact, was not a partition but a disannexation, as Poland preserved its statehood, was one of the most just acts in world history. It ended the evils of four centuries, bringing a conclusion to four centuries of oppression.

While the partition restored justice, those involved in its restoration decided to proceed too far. The agony of the Polish state in the late 18th century fed the undeniable desire of its neighbours to profit from what lay unguarded in front of them (this was similar to the feeling that the paralysis of the Russian state in the 13th and 14th centuries created in the hearts of the Polish and Lithuanian kings). The result was the final partition of Poland—the execution of a whole nation—which led to the forceful inclusion of a hostile state element into the Russian state organism. It did not take too long for the consequences of this action to be realised: the Polish uprisings against the Russians, who occupied Warsaw, were as understandable and justifiable as the Russian uprising against the Polish who had occupied the Kremlin a few centuries before. The troops of Skrzynecki[3] and the *kosynierzy*[4] of Sierakowski[5] were in exactly the same situation fighting the Russians as the militiamen of Pozharsky[6] and the Cossacks of Khmelnytsky[7] when they fought against the Polish.

---

[3] Jan Zygmunt Skrzynecki was a Polish general, commander-in-chief of the 1830–31 November Uprising against the Russian Empire.

[4] *Kosynierzy* is the Polish term for soldiers (often peasants) armed with war scythes.

[5] Zygmunt Sierakowski was a captain in the Russian Imperial General Staff. Born in the Volynskaya Guberniya (formerly Polish territory), he was a Polish patriot. After the beginning of the 1863–64 January Uprising against the Russian Empire he took leave of absence from service and became one of the leaders of the uprising.

[6] Dmitry Pozharsky was a Russian prince who formed the Second Volunteer Army during the 1611–12 Polish–Muscovite War against the Polish–Lithuanian Commonwealth's occupation of Russia.

[7] Zynoviy Bohdan Khmelnytsky was a Ukrainian hetman who led the 1648–57 Khmelnytsky Uprising against the Polish–Lithuanian Commonwealth.

The next period was marked by the decline of the Russian state, the revival of Poland, and the subsequent reappearance of the unhealthy desire to grab everything poorly guarded. The result was the 1921 Peace of Riga and the subsequent reannexation of the 'dissidents'.

As we can see, in the centuries-long Russo-Polish confrontation, the initial injustice was committed by the Poles. This, however, should not serve as justification for all subsequent actions by the Russian side. The Warsaw Governorate within the Russian Empire was as unjust as the Volhynian Voivodeship within the Polish–Lithuanian Commonwealth. While harmony between Russia and Poland was restored for a short period of two decades (1772–94), both sides could not and did not want to maintain it. Justice had always migrated from one camp to the other, though there was a certain advantage on the Russian side, as 'the original sin' was committed by the Polish.

The same analysis can be carried out in other specific cases—for example, in the confrontation between the Russian and German peoples. The root of this conflict goes back to the Livonian Brothers of the Sword, which by fire and sword exterminated Slavic tribes in the name of militaristic Germanism and caused significant damage to the northern regions of the Novgorod Republic (despite the 1240 Battle of Neva and the 1242 Battle on the Ice). From the Battle on the Ice, through the 1558–83 Livonian War, the 1709 Battle of Poltava, the 1714 Battle of Gangut, [...], the 1915 Gorlice–Tarnów Offensive, to the 1918 Treaty of Brest-Litovsk, justice had always been on the Russian side (with the single exception of the Seven Years' War).

In analysing the Franco-German conflict, the starting point should be regarded as 1806—the Battle of Jena-Auerstedt, which was followed by the Treaties of Tilsit (the prototype of the authoritarianism of the Treaty of Versailles). The three-centuries-long struggle between the Bourbons and the Habsburgs never had a nationalistic pattern, let alone an ethnic one. While Prussia should be blamed for initiating this confrontation, a significant contribution to the conflict was made by Napoleon's inability to contain himself and the utopia of the barbarians of 1789, who promoted the 'principle of nationalism' (initially as a countermeasure against tyrants, then as a self-sufficient principle). Scattered around the continent, their seed of nationalism found the most fruitful soil in Germany. Owing to these theories of nationalism, the Germans of twenty-six separate states realised, for the first time, that they belonged to a single nation. Owing to these theories of nationalism, Fichte[8] was able to write his 'Addresses to the German Nation'—something he could not have done fifteen years before, given the absence of this German nation. The creation of the German nation occurred in the period between 1806 and 1813. Its doctrine was conceived in the first three decades of the 19th century (Fichte, Hegel, etc.). The process of perfecting this doctrine, which lasted for another hundred years, led to the wars of 1870 and 1914—wars in which justice was undeniably on the French side (the 'Ems Dispatch' fabricated by Otto von Bismarck in 1870 and the false German allegations that several French aircraft bombed Nuremberg in 1914).

Restricting ourselves to these examples, we will turn to a discussion of the aims of war.

* * *

The greatest barbarian of the 19th century—Clausewitz—proposed the theory of an 'integrated war' of annihilation. This theory of Clausewitz was best implemented by his most ardent student, Lenin. This is why we are going to call it 'Clausewitzism-Leninism'. Its main idea is based on annihilation and extermination of the enemy: not only the defeat of the enemy's armed forces, but the full enslavement and extermination of the enemy as a nation (according to Clausewitz and his followers) or as a class (according to Lenin).

This misanthropic theory was implemented (though quite gingerly) by the Germans during the Great War: atrocities in occupied territories, their use of hostages and terror, asphyxiating gases, unrestricted submarine warfare, and support of an internal enemy to subvert the enemy state. It was also implemented by the Bolsheviks, though on a much greater scale.

We should completely reject this Clausewitzian false doctrine, as well as its derivative, 'Leninism'. This false doctrine corresponds to neither Christian Orthodox values, nor Russian historical tradition, nor Russian military ethics, nor simple common sense.

---

[8] Johann Gottlieb Fichte (1762–1814) was a German philosopher who became a founding figure in the philosophical movement known as German idealism.

Wars are not waged to kill—they are waged to win. While the immediate aim of war is victory, its final aim is peace and restoration of harmony, which is the natural state of human society.

Everything else is disastrous brutality. In dictating peace to a defeated enemy, one should be driven by modest aims and not drive him to despair by excessive demands. Such demands can only lead to hatred and, therefore, sooner or later, to new wars. Instead, a victor should compel the enemy to respect him. However, in doing so he must avoid chauvinism and should respect the national and simple human dignity of the defeated.

There is no higher aim in politics than 'peace on earth, goodwill to all men'.[9] This ideal, to which politics should strive, is inconsistent with a nation of helots enchained by Clausewitz and his followers, or with Lenin's idea of turning the universe into a graveyard.

Chapter 3: War and Peace

Peace is the natural state of humanity, as peaceful existence fosters its spiritual development and material wealth. War, for humanity, is a phenomenon similar to the illness of human organism.

Therefore, war is a pathological phenomenon that disturbs the proper metabolism of the state organism. The organism of a nation at war can be compared to the human organism during a period of illness. The main difference between these two is that a human organism does not decide to get ill, while state organisms consciously take the risk of 'military illness'.

Many wars were of great service to humanity. The Roman campaigns subjugated the whole ancient world, bringing civilisation, law and, later, Christianity to the Iberian and Celtic tribes. The javelins of Roman legionnaires became the pillars of the European state (the Slavic tribes, conquered by the Germans, were characterised by anarchy and weak statehood exactly because they lacked this Roman influence). In the time of the crusades, the West borrowed from the East its sciences, and thus the semi-barbarian Westerners learnt a lot from the enlight-

---

[9] This is from the Orthodox Church version of the *Gloria in excelsis Deo*, based on Luke 2:14.

ened Arabs. At the turn of the 16th century, the Italian campaigns of Charles VIII and Louis XII brought the Renaissance to France. In return, France gave the Renaissance the splendour and magnificence of Europe, something that its motherland of Italy, divided into many city-states, could not give.

Generally speaking, while war itself should always be considered a calamity, its consequences can sometimes be salutary. The 1914–18 War was a calamity on a scale which can very rarely be seen in human history. The Russian catastrophe of 1917 can be matched only by the Black Death of the 14th century or (to a lesser degree) by the invasion of the Mongols.

However, the Russian Revolution was not the child of war, but its adoptee. This revolution was the daughter of the 19th century and its outdated theories. The real child of the Great War was Fascism and its associated ideas, which opened new horizons before humanity. It gave humanity new forms of social order, leading human thought and society out of the dead end to which they had been pushed by the barbarians of 1789 and their followers (the materialists of the 19th century). This positive spiritual consequence in itself allows us to accept that ten million people did not give their lives for nothing.

* * *

Regardless of all that, war is a great and undeniable evil. The decision to use this evil (this illness) should be taken only in desperate situations, when fighting fire with fire is the last resort and all other arguments have failed. Generally speaking, a bad peace is better than a good fight. Any exception to this rule can be considered only when the peace is so bad that it threatens to damage the morale and well-being of the state.

The state organ responsible for making these decisions is called diplomacy, and it consists of the central administration and its representatives abroad. There are two different schools of diplomacy.

The first one is old (cabinet) diplomacy. In this case, state affairs are entrusted to people specially assigned to this job, people who were raised, educated and, one might say, born for this task. The very process of climbing the ladder of the service guarantees the competence of these people. Before they reach the rank of ambassador or special envoy (equivalent to a general), each of them spends time in four or five capitals, studying practically four or five states, their rulers and other diplomats working in these states (i.e., a large portion of his foreign colleagues).

The pathological era of 1914–18 was followed by a period of collective softening of the brain (the consequence of the serious concussion caused to the world by this war). This period is called the era of 'democracy'. The foreign policy consequence of this epoch was the replacement of the old cabinet school of diplomacy by a new one—the school of carnival. The professionals of state affairs were replaced by amateurs. Knowledgeable people were replaced by ignorant speakers used to addressing rallies, who held the title of 'people's representatives', but did not always have even a primary school certificate.

The results of this newly minted diplomacy of the unwashed were not long in coming. It brought the 'spirit of internal politics' with its atmosphere of rallies and backroom deals into international relations. In comparison with professional diplomats—people who are not (generally) tied politically—the hands and feet of the political leaders of democracy are tightly tied. Any internal political intrigue with no meaningful connection, any minor incident inside the party, forces them to hastily leave international conferences, where questions of paramount importance are being debated (with greater or lesser degrees of incompetence). This unproductive break leads to an undefined, stressful and nerve-racking situation of some weeks until some untimely self-dissolved provincial Masonic lodge or parliamentary faction (a grain of sand that stopped the train) is calmed down. In the era of democracy, any international problems are seen, first and foremost (and sometimes exclusively), from the perspective of internal politics (i.e., party interests). Personal success in parliament or during a political campaign is the sole concern of all these 'carnival-style' Talleyrands.[10]

---

[10] Kersnovski refers to Charles Maurice de Talleyrand-Périgord, a French diplomat whose career spanned the reign of Louis XVI, the years of the French Revolution, and the reigns of Napoleon, Louis XVIII and Louis-Philippe; his survival turned the name 'Talleyrand' into a byword for crafty, cynical diplomacy.

ANTON KERSNOVSKI

The scrupulous 'jeweller-style' work of professional diplomats is much above the skills of carnival-style diplomats (the leaders of the masses), whose positions, words and actions are focused on the present and designed for the intellect of the masses. Affairs previously discussed by competent people in a calm and productive setting are now argued in a blazing atmosphere before the masses (masses that become demoniac at the first sight of gladiators' blood in the arena).

In support of this point, it is enough to recall the inglorious end of the League of Nations. The bankruptcy of this institution, as well as of the idea that spawned it, is so obvious that it releases us from the need to prove it by numerous facts.

The old diplomacy is frequently accused of 'provoking' wars. This accusation, however, can be made only by people who consciously quarrel with history (the most undemocratic science). The history of the last two hundred years teaches us that for every war 'provoked' by diplomacy (provoked, we should not forget, by order of the corresponding governments) there were three wars that the old cabinet diplomacy failed to prevent (though the new carnival-style diplomacy will fail too, as there are no diplomatic notes that will stop a tiger that has already decided to jump). And for every war that could not be prevented, there were at least ten wars that did not occur owing to the timely, tactful, proper and (to the savage and ignorant masses) invisible intervention by professional diplomacy.

Chapter 4: On Disarmament

In the time of plague pandemics during the Middle Ages, as well as in the more modern time of the Cholera Riots, the mob attacked healers and doctors, seeing the extermination of physicians, who allegedly spread the infection, as a means of getting rid of the trouble.

The 'intellectual mob' of the 20th century (the so-called pacifists), as well as the carnival diplomacy that directs them (and is simultaneously led by them), sees the dissolution of the military as a means to get rid of wars. According to them, the presence of armed forces is the very cause of evil: status-seeking, bloodthirsty generals and profit-pursuing armaments manufacturers drive the country into a military adventure with the connivance of 'degenerate dynasties' and 'secret' (and, worse, entitled) diplomacy.

STRATEGIYA

This is the threadbare pattern of 'democratic pacifist' thinking. This viewpoint of the orator preaching on a barrel in front of the ignorant mob has become the official doctrine of the 'advanced democracies' of the 1920s and 1930s—the era of democratic lunacy. According to this point of view, to avoid illnesses we should beat our doctors and shut down pharmacies; to avoid fires we should dissolve fire brigades; and to avoid train derailments we should abolish semaphores and signalmen.

The idea of disarmament has been preached in two different ways. The first implies the replacement of regular armies with people's militias (i.e., replacing 'militant professionals' with allegedly peace-loving militias). The second suggests that the renunciation of several technical means and advantages (e.g., military aviation or some types of it, chemical gases, heavy artillery) would make wars 'less bloody'.

The myth of a 'peace-loving militia', which was the obsessive idea of Jaurès,[11] is old and hackneyed, similar to any ideology of 1789 and 1848. The most undemocratic science—history (a science most hated by democracy, because for democracy history means death)—teaches us that the opposite is true. It is enough to recall the people's militias of antiquity: the Cimbri, the Teutons, the Huns, the Mongols. In the early modern period, the 'peace-loving' militia revelled in feasts on the blood of either Protestants or Catholics (depending on its own religion). In the late modern period, the armed citizens of the Republic 'declared peace with the world', pushing 'liberté, égalité' across Europe with bayonets and gunpowder, expanding the borders of France to the Rhine, and introducing 'peace' to the people of the Netherlands, Switzerland and Italy by the establishment of the Batavian, Helvetic, Cisalpine and Parthenopean republics with guillotines on every square.

Similarly unconvincing, though less hackneyed (as it was born not in 1789, but in 1919), is the idea of promoting disarmament by the 'withdrawal from service' of several battle-tested military means.

Asphyxiating gases have been prohibited. However, neither chemistry as a scientific subject, nor chemical laboratories, nor medical institutions, nor manufacturers of artificial fertilisers or paint, nor

---

[11] Jean Jaurès (1859–1914) was a French anti-militarist, the leader of the French Socialist Party, and one of the main historical figures of the French Left.

common pharmacies have been prohibited. While none of these can be prohibited, all these institutions can be easily and quickly modified for the production of asphyxiating gases. While there are attempts to prohibit military aviation, no one speaks of prohibiting civilian aviation (as it cannot be prohibited), even though all mail and sport aircrafts can easily be turned into bombers and fighters. While there are attempts to prohibit tanks, no one speaks of prohibiting automobile and tractor production. While there are attempts to prohibit heavy artillery, no one speaks of prohibiting the metallurgy industry. All these attempts at disarmament (voluntary, semi-voluntary, or compulsory) seem to recall a good deal of the laws aimed at compulsory sobriety, which were so unsuccessful in both Russia and the United States. In the case of disarmament, they attempt to prohibit 'military chemistry' without prohibiting the chemical industry; in the case of the dry laws, they attempt to prohibit vodka without prohibiting moonshine (as no one has the power to prohibit it) or the means to buy counterfeit wine of every kind. Both these ideologies (disarmament and dry laws) are based on a complete misunderstanding of human nature and human psychology, and therefore, having no chance of survival, they are both doomed to fail.

* * *

Technological means, strange as it may seem, do not increase the bloodshed of war. No battle of the Great War reached the level of carnage of the 1812 Battle of Borodino, which was fought with flintlock muskets (at Borodino 100,000 people were killed on both sides in eight hours; at Verdun 700,000 though in eight months). The main impact of technology on tactics is manifested in the protracted nature of operations, increased stress and, in general, less bloodshed. During the seven hours of the 1870 Battle of Gravelotte, the same number of people were killed as during a week (of hard work by magazine rifles and rapid-fire artillery) of the 1904 Battle of Liaoyang.

The bloodshed of a battle is not the result of 'technology', but of bad tactics, the 'rhythm' of operations, the quality of the troops and fierceness of the fighters [...]. The carnage of barbarian tribes, armed with stone axes, was significantly bloodier than contemporary fire-based battles. In 1914 the French military, armed with relatively primitive technology, lost on average 60,000 killed and wounded every month (this was their price for bad tactics). In 1918, when the front was flooded by 'deadly technology', the casualties did not increase (as some might assume) but, on the contrary, were three times lower (on average 20,000 per month).

'Technology', therefore, tends to decrease casualties, and not increase them. Asphyxiating gases, regardless of their undeniably 'vile' nature, kill on average only two people out of every hundred affected. However, the 'humane' bullet kills a quarter of those it successfully hits.

This convincingly demonstrates the incompetence of the 'technological disarmament' theory. Any restriction of technology by 'disarmament' would not lead to less bloodshed on the battlefield.

The idea of 'moral disarmament' has more (though not much more) substance in it. This idea was the most beloved argument of Geneva's snobs in the late 1920s and early 1930s. They just forgot that to achieve moral disarmament among nations, they needed first and foremost to prohibit the very root of conflicts among these nations—politics. However, to prohibit politics, they would need to prohibit the very reason for the existence of politics—the uninterrupted advance of human society with its spiritual, intellectual, material and physical developments. [...] Such 'moral disarmament' would definitely eliminate conflicts, but it would also eliminate the reason for their existence—life itself.

There is only one category of people who are safe from all diseases—the dead. Only extinct humanity will be safe from its main disease—war.

* * *

Thus, if we want to protect the state organism from the pathological phenomenon called war, then infecting it with pacifist ideas is the last thing we should do. If we desire our organism to resist pathogens, we should not weaken it, hoping that the guilty conscience of microbes moved by our vulnerability will prevent them from attacking a weakened organism. Instead we should reinforce it as much as we can. By strengthening our state organism by an appropriate programme of treatment (external and internal) and prophylaxis, we will increase its resistance to the pacifist utopias from outside, as well as Marxist false doctrines from inside, thus reducing the risk of both external (interstate) and internal (civil) wars.

Only weak states are attacked. No one attacks a strong state. Weak states that appear strong are attacked less frequently than strong states that fail to demonstrate their strength in a timely manner, thus appearing weak from the outside.

The famous 1888 Schnaebele affair[12] almost led to a new Franco-German war. However, Bismarck backed down at the last moment: the French Army had been recently equipped with the new Lebel Model rifles with magazines, while the German Army still used single-shot rifles. In the 1880s, France held dreams of revanche and was strong (the Dreyfus affair, which seriously poisoned its body, occurred much later), and mighty Russia of Alexander the Peacemaker was threatening from the east. Bismarck decided to delay his adventure.

Another example is the 1904 Russo-Japanese War. Prince Ito[13] proved himself more decisive than Bismarck in his time. Would the Japanese have dared to attack us if the Port Arthur squadron had appropriate docks, if in place of Kashtalinskii's[14] division at the Yalu River[15] there had been three or four corps, and if Manchuria had been connected to Russia not by a single-track (and unfinished) railway, but by a fully completed quadruple-track railway? Would the Japanese have dared to attack if they knew that the Russian state was not lulled by the opiate originating in The Hague, and Russian society would not send welcoming telegrams to the Mikado but would defend the interests of its country?

Another example is the 1914 Great War. Germany initiated a war because it did not want to deal with a much stronger Russian army in 1920, the likely outcome of the seven-year-long 'Great Army

---

[12] The Schnaebele Affair took place in April 1887 and not in 1888 as Kersnovski states.
[13] Prince Itō Hirobumi was a senior Japanese politician, and president of the Privy Council of Japan during the Russo-Japanese War.
[14] Nikolai Kashtalinskii was a Russian Imperial general, commander of the Third East Siberian Infantry Division during the Russo-Japanese War.
[15] Kersnovski refers to the 1904 Battle of the Yalu River.

STRATEGIYA

Programme' approved in 1913.[16] Simultaneously, Germany declared war on France: the timidity and frivolity of France's leaders (Viviani's[17] order for a ten-kilometre withdrawal of the French Army from the border, as a display of peaceableness) were perceived by Germany as proof of France's weakness. This was a typical example of someone attacking a strong state because it seemed weak.

An analysis of all wars and conflicts, from the distant past to nowadays, proves the correctness of the rule used by Yermolov[18] during the conquest of the Caucasus, though it was only coined a hundred years later by Lyautey[19] in Morocco: 'Show your strength to avoid using it.' This rule should lie at the heart of any healthy politics.

Generally speaking, it is important to remember that humanity has paid more in blood to 'ideologists' than to conquerors—more blood was spilt by the followers of Rousseau's utopias than by the hordes of Tamerlane.

Chapter 5: The Nature of Military Affairs: Military Art and Military Science

Do military affairs belong to the field of science or art? To answer this question, we need to take into consideration the dual nature of military affairs.

Military affairs consist of two elements. The first is rational, measurable and physical. It is given to precise analysis and classification. The

---

[16] Kersnovski refers to a reform led by the Russian Minister of War, Vladimir Sukhomlinov, but confuses the dates. On 10 July 1913 Sukhomlinov approved the 'Small Military Programme' and on 24 June 1914 the 'Great Military Programme'. Both reforms were to be completed by the end of 1917 and involved a significant rearmament and reorganisation of the armed forces.

[17] Jean Raphaël Adrien René Viviani was the French Prime Minister for the first year of the First World War.

[18] Aleksey Petrovich Yermolov was a Russian Imperial general who commanded Russian troops in the Caucasian War between 1816 and 1827.

[19] Louis Hubert Gonzalve Lyautey was a French general and the first French Resident-General in Morocco from 1912 to 1925.

second is irrational, spiritual and unmeasurable—something that Napoleon called 'la partie sublime de l'art'.

The rational, physical part of military affairs belongs to military science. The irrational, spiritual part belongs to military art. Any sighted person can see though his bodily eyes, but not everyone can see through his spiritual inner eye. Art is given by God, while science is learned through human labour. Zeus could be sculpted only by Phidias, while an anatomical scheme of the human body can be prepared by any medical student.

As a general rule, art (the destiny of a chosen few) is higher than science (the destiny of the rest). It is important to note, however, that when science reaches its highest levels, it is also marked by genius—it has its own 'partie sublime'. Mendeleev and Pasteur ennoble the human race to the same degree as Dostoyevsky and Goethe.

Like precious metals, art cannot be used in its pure form. Like any alloy, it should also include a certain portion of science.

When a composer experiences inspiration and his soul plays soundless chords, this is a moment of pure art, so-called absolute art. When he takes a pen and starts to transfer his inspiration onto paper (otherwise it would be lost to him and the rest of humanity), this is the moment when art is mixed with the alloy of science. A composer must know notes, beats, counterpoints, and be able to read sheet music; a poet must know grammar; a sculptor must know human and animal anatomy as well as the qualities of plaster, bronze and marble.

The same principle applies to military art. Even the most ingenious plan risks turning into a chimera if it does not conform to the rules of reality. Even the greatest general has no excuse for neglecting the elements of military science, even though he himself perfects this science and conforms (often instinctively) to its principles.

The higher the percentage of the noble metal in an alloy, the more precious this alloy. The more the irrational element of art (in a commander's actions) dominates the rational element of science, the higher is his generalship. Napoleon, in the majority of his operations, and Suvorov, in all his operations, offered us twenty-three-carat gold. The generalship of Frederick the Great (greatly tarnished by routine and 'methods') was fourteen-carat gold. While the generalship of Moltke the Elder (a talented general but not a genius) offered us silver of quite a high degree of fineness, the generalship of his nephew was nothing more than a cheap tin alloy.

* * *

Military science should be subordinated to military art. In other words, military art should come first, and only then military science.

There are some cases where science should cover or replace art (or, following our previous metaphor, science should play the role of rolled gold). These cases correspond to the most critical periods of military art, epochs when the art is in decline, when its soul leaves the already obsolete though still existing forms and seeks new ways of existence. This was the case in the West during the second half of the 17th century, when the armies of mercenaries sought salvation in the routine of lines of battle and the sophisms of the 'five-march system'.[20] New ways were found by the French (during the so-called Great Revolution, which armed the whole nation), leading to the revival of military art during the Revolutionary and Napoleonic Wars. The same occurred during the 1914–18 War, a war that saw both the culmination of armed nations ('hordes') and their decline. After the war, salvation for military art was found in the shape of an old Russian system that combines the idea of quantity (*Narodnoe Opolchenie*—people's militia) with the idea of quality (*Druzhina*—professional army). The last time this system had been implemented was in 1812 (Rostopchin[21] was responsible for the former and Kutuzov[22] for the latter). Nowadays, this old

---

[20] The 'five-march system' is a military logistics concept that implies that an army should not proceed further from its supply depots than five marches at most. If it becomes necessary to proceed further, a fresh line of supply depots should be formed.

[21] Count Fyodor Rostopchin was a Russian statesman who, during the French invasion of Russia in 1812, was responsible for the defence of Moscow and took every means available to rouse the population to defend the city.

[22] Prince Mikhail Golenishchev-Kutuzov was a field marshal of the Russian Empire. He served as one of the finest military officers and diplomats of Russia under the reign of three Romanov tsars: Catherine II, Paul I and Alexander I. Kutuzov is considered to have been one of the best Russian generals. During the French invasion of 1812, he insisted on abandoning Moscow to save the remains of the Russian Army.

Russian system is called 'Seeckt's system'²³ by foreigners (whose incompetence can be excused) and Russian ignoramuses (whose ignorance is unforgivable).

The Great War was the period of the 'twilight of military art'—a period described by Foch²⁴ as showing 'the forced absence of sufficient genius' ('l'absence forcée d'un genie suffisant'). Salvation in this case was found in a collective, the best example of which was the General Staff of the German Army.

The burden of generalship in the Great War (a war that combined a lot of scientific alloy with too little gold of art) fell on these collectives, which existed by this time in every army, and on their most distinguished members. This is why 'grey' is the best colour that characterises generalship during this war. There were few exceptions, however, such as the works of art created by General Yudenich²⁵ during the Caucasus Campaign, the battles of General Mangin²⁶ (who deserves to be called a French Skobelev),²⁷ some operations of Hindenburg on the Eastern Front, Kluck's performance on the Ourcq River,²⁸ and several others.

There was not much art, and it was concentrated in the works of only a few leaders. In the most decisive moments, the creations of Joffre,²⁹

---

[23] Kersnovski refers to the way Hans von Seeckt, the de facto chief of the Reichswehr, organised the armed forces of the Weimar Republic.

[24] Ferdinand Foch was a French general and military theorist who served as Supreme Allied Commander during the First World War.

[25] Nikolai Yudenich was a Russian Imperial general who served as chief of staff of the Russian Caucasus Army during the First World War. He was a leader of the anti-communist White Movement in Northwestern Russia during the Civil War.

[26] Charles Emmanuel Marie Mangin was a French general known for his front-style command during the First World War.

[27] Mikhail Skobelev was a Russian Imperial general famous for his heroism during the 1877–78 Russo-Turkish War.

[28] Kersnovski refers to the aggressive decision of Alexander von Kluck, the commander of the German First Army, to advance towards Paris at the beginning of the First World War, against the orders of his superiors.

[29] Joseph Joffre was a French general who served as commander-in-chief of French forces on the Western Front at the beginning of the First World War.

STRATEGIYA

Gallieni,[30] Foch and Mangin (who together made up the famous 'Le quadrilatère') surpassed those of Moltke the Younger, Kluck, Falkenhayn[31] and Ludendorff, in the same way that Hindenburg, Ludendorff, Falkenhayn and Mackensen[32] surpassed the Grand Duke,[33] Zhilinsky,[34] Ruzsky[35] and Ivanov.[36] This situation defined the character of war, predetermining its outcome, even though the German collective with its *Durchschnitt*,[37] its scientific basis, its formulation and development of doctrine, and its establishment of the rational part of military affairs was much better than the French. Personalities, as usual, played the most decisive role. Military art, after all, is a personal quality. Though the French military art was not of the best quality (due to reasons beyond the control of the leaders), the achievements of the French collective proved to be better than those of the scientific rationalist Germans.

Science blends with art only within ingenious personalities. When it comes to the field of generalship, military science (especially when

---

[30] Joseph Gallieni was a French general. From October 1915 to March 1916 he served as Minister of War.

[31] Erich von Falkenhayn was a German general. He was the second chief of the German General Staff from September 1914 until 29 August 1916.

[32] August von Mackensen was a German general. He was considered one of the most prominent and competent military leaders during the First World War.

[33] Historically, in Imperial Russia, the title 'Grand Prince' was a generic title for members of the Imperial family. Here, Kersnovski refers to Grand Duke Nicholas Nikolaevich of Russia (the Younger) who was commander-in-chief of the Russian Imperial Army in the first year of the war.

[34] Yakov Zhilinsky was a Russian Imperial general. He served as chief of staff from 2 February 1911 to 4 March 1914. At the beginning of the First World War, he assumed command of the Northwestern Front. After the fiasco of the East Prussian Campaign, he was relieved of his command.

[35] Nikolai Ruzsky was a Russian Imperial general. He was the commander-in-chief of the Northwestern Front during the First World War.

[36] Nikolai Ivanov was a Russian Imperial general. He was the commander-in-chief of the Southwestern Front during the First World War.

[37] Kersnovski uses the German word *Durchschnitt*, which literally translates as 'average'.

it attempts to compensate for the lack of art) leads to quite rudimentary outcomes. Purely scientific generalship (either with a very weak artistic element or without it) can be compared to the mathematical calculation of a circle's circumference. While science offers the number π, allowing one to make highly accurate calculations, it fails to provide the means to comprehend the 'irrationality' of the circle. A scientific 'methodology' may get close to the intuition of art, but it cannot compete with it. An invisible but tangible barrier will always separate these two—while Salieri 'checked [his] harmony by algebra', he did not reach the heights of Mozart.

Art's superiority over science is similar to the superiority of spirit over mind, individual personality over the masses, soul over matter.

* * *

Military art, like any other art, is national, as it represents the spiritual creativeness and background of the nation that developed this art. We can easily distinguish between the Russian, French, Italian, Flemish and other schools of painting. We can easily distinguish the enchanting sounds of Russian music from any foreign one. The same applies also to the military field: 'The Science of Victory'[38] could be created only by a Russian genius; 'On War' could be written only by a German.

Of all existing arts, only two—military and literary—serve as sensitive barometers of national self-consciousness. While the extent of their reactions to the changes in national self-consciousness is similar, the nature of these reactions is different. Military art is organically connected to national self-consciousness and therefore grows and declines together with it. Literary art, which is less dependent on national self-consciousness (as the connection between them is less organic), reacts differently. It reflects the fluctuations of national self-consciousness in its mirror. While it transforms the 'material', the character of the change remains more or less the same. Lomonosov, Pushkin and Chekov—three artists, the first of whom mirrors the sunrise of Peter's empire, the second its noon, and the third its sunset.

---

[38] A military manual written by Alexander Suvorov (1720–1800), one of the most prominent figures in Russian military history.

STRATEGIYA

Analysing these 'barometers' is a very interesting affair, as military affairs represent the synthesis of a nation's 'actions' and literature represents the synthesis of its 'words'. The genius of Rumyantsev[39] corresponds to the genius of Lomonosov.[40] Suvorov goes hand in hand with the soaring eagle-like Derzhavin.[41] The generation of the heroes of 1812 (the shining generation of Bagration[42] and Denis Davydov)[43] produced 'the poet in the land of soldiers'—Zhukovsky.[44] The younger representatives of this generation were Pushkin and Lermontov. The epoch of Alexander the Liberator gave us the luminaries of Russian self-consciousness: Dostoyevsky and Aksakov (literature) and Skobelev (military). Then we can see a clear decline: between the twilight of the setting 19th century and the pale rise of the 20th century we can see the dim figures of Kuropatkin[45] and Chekov.

* * *

Art, therefore, is a national affair. Regardless of the field of art (whether it is literature, painting or military affairs), nationality is its most distinctive aspect, its so-called 'flavour', its quintessence. An objective, international, 'global' art does not exist.

Science, however, is a different story. While nations' spirits differ one from another (and, therefore, their arts, which are the creations

---

[39] Pyotr Rumyantsev (1725–96) was one of the foremost Russian generals of the 18th century.

[40] Mikhail Lomonosov (1711–65) was a Russian polymath, scientist and writer who made important contributions to literature, education and science.

[41] Gavriil Derzhavin (1743–1816) was one of the most highly esteemed Russian poets before Alexander Pushkin.

[42] Pyotr Bagration (1765–1812) was a Russian general and prince of Georgian origin, prominent during the Napoleonic Wars.

[43] Denis Davydov (1784–1839) was a Russian soldier-poet of the Napoleonic Wars who invented the genre of Hussar poetry.

[44] Vasily Zhukovsky (1783–1852) was the foremost Russian poet of the 1810s and a leading figure in Russian literature in the first half of the 19th century.

[45] Aleksey Kuropatkin was the Russian Imperial Minister of War from 1898 to 1904. He is often held responsible for major Russian defeats in the Russo-Japanese War.

of their spirits), the differences between their intellectual lives (between the groups of thinking people—'elites') are much smaller. In science, unlike in art, there are many more 'contiguous points'.

Mathematics, physics, chemistry, medicine—all are objective sciences, as is the dogmatic part of philosophy. Everyone solves the Pythagorean theorem in the same way, regardless of whether he is French or German, a Communist or an anarchist.

The empirical parts of philosophy, history, sociology, law are social sciences and therefore are national and subjective, as they deal with researching the phenomena of nations' lives, making conclusions regarding their development. While a French and a Russian will calculate the Pythagorean theorem in the same way, they will describe the campaign of 1812 differently. Moreover, the interpretation of phenomena on which these sciences focus depends not only on the nationality of the interpreter, but also on his subjective political worldview. A comparison between Ilovaysky[46] and Milyukov,[47] or between Gaxotte[48] and Lavisse,[49] serves as a good example of that. By adopting the Soviet method of 'historical materialism' and the 'class approach', it is possible, for example, to turn Pugachev's 'general' Khlopusha[50] into a central figure in Russian history, allocating to him two hundred pages and leaving half a page for Rurik,[51] Ivan the Terrible and Peter I combined.

Military science belongs to the category of social sciences. Therefore, it is national and subjective. While some consider it a part of

---

[46] Dmitry Ilovaysky (1832–1920) was a Russian historian who wrote a number of standard history textbooks.
[47] Pavel Milyukov (1859–1943) was a Russian historian and liberal politician.
[48] Pierre Gaxotte (1895–1982) was a French historian.
[49] Ernest Lavisse (1842–1922) was a French historian.
[50] Khlopusha (real name, Afanasii Sokolov) (1714–74) was a Russian peasant and one of the closest associates of Yemelyan Pugachev, the leader of 1773–75 Pugachev Rebellion in the Russian Empire after Catherine II seized power in 1762.
[51] Rurik (830–79), according to the 12th century *Primary Chronicle*, was a Varangian chieftain of the Rus. This legendary figure was considered by later rulers to be the founder of the Rurik dynasty, which ruled the Kievan Rus and its successor states.

sociology, in our humble opinion this is a mistake. *Military science is sociology in itself*, as it embraces the whole complex of social disciplines. *It is, however, pathological sociology.*

Peace is a natural state of humanity, and sociology researches the phenomena of this natural state. War is a pathological phenomenon that can be characterised as an illness. The nature of an ill organism, its characteristics and functions are not similar to those of a healthy organism, and looking at both through a similar prism is impossible.

Therefore, military science is a sociology of war or (considering war as a pathological phenomenon) pathological sociology. [...]

PART TWO: ON ELEMENTS OF WAR

Chapter 6: Politics and Strategy

Politics guides the nation and state. Strategy guides the armed part of the nation, a very specific emanation of the state, which is called the army.

Politics is a whole, whereas strategy is a part of it. Strategy works in a field bounded by politics. It represents the politics of war, while war itself is an element of state politics. Hence, strategy is one of the elements of politics, one of its most important elements.

The aim of politics is to prepare the stage for strategy, to place it in the most favourable situation at the beginning of a war, to ease its work during the war, and to reap (as best as possible) its fruits after the war.

Diplomacy and strategy are the two hands of politics. And it is absolutely necessary that the right hand should always know what the left does, and vice versa. Before taking any substantial decision of national and, even more so, international significance, the politician should turn to the strategist and ask him: 'I intend to do so-and-so; are we powerful enough for that?' If the strategist's answer is positive, then the politician can raise the national flag and step boldly into the international arena. However, if the strategist's answer is negative, then the politician can do nothing else but furl the banners, draw in his horns and lower his voice, sacrificing sometimes his nation's pride to avoid the worst-case scenario. Moreover, in this type of situation, the strategist's duty is to warn the politician, without waiting to be asked.

Before Austro-Hungary decided to annex Bosnia and Herzegovina in the winter of 1909, Aehrenthal[52] consulted Hötzendorf.[53] Only after hearing that the Russian military was disorganised after the Russo-Japanese War, and that their own military was strong enough, did Aehrenthal dare to make this decisive move. Simultaneously, Izvolsky[54] asked General Roediger:[55] is Russia capable of reacting to this situation to protect its honour as a great power? He received an honest, straightforward and simple answer. While Russia was saved from catastrophe, it paid the high price of humiliation for this act of salvation.

A classic case of cooperation between politics and strategy (when the politician turns to the strategist) occurred in the first half of July 1870, at the beginning of the Franco-Prussian conflict regarding the candidacy of a Hohenzollern for the throne of Spain. At this time, King Wilhelm I was enjoying time at the Ems spa resort and, resting on the laurels of the 1864 Second Schleswig War and the 1866 Austro-Prussian War, he was in a peaceful mood. Bismarck, however, saw in a war with France the final stage of the unification of Germany—the monumental goal which his politics had continuously tried to achieve.

On 16 July, Bismarck, Moltke and Roon were having breakfast in Ems when the chancellor received a copy of a telegram sent to the French ambassador in Berlin. This was the reply of the French government to a Prussian statement, written in very pacific and mild language. All three became very depressed, realising that with their peaceful king, a war would be impossible and the unification of Germany would have to wait, if not for the Greek calends, then at least for a very long period of time.

Bismarck stood up and made a decision. 'Tell me,' he asked Roon, 'does our army have everything it requires?' Roon replied: 'Absolutely.'

---

[52] Alois Lexa von Aehrenthal was Foreign Minister of Austro-Hungary between 1906 and 1912.
[53] Franz Xaver Conrad von Hötzendorf was a field marshal and chief of the General Staff of the Austro-Hungarian Army and Navy from 1906 to 1917.
[54] Count Alexander Izvolsky was Foreign Minister of the Russian Empire between 1906 and 1910.
[55] Alexander Roediger was the Minister of War of the Russian Empire from 1905 to 1909.

Bismarck turned to Moltke: 'Can you guarantee a successful course of war?' Moltke said: 'I can guarantee that.'

'Then,' Bismarck writes in his memoirs, 'I went to another room, sat at the table and altered the whole text of the French telegram, changing its tone and content and replacing pacific sentences with insults.' In other words, he falsified the telegram and submitted it to the king. King Wilhelm I, indignant at the French 'impudence', abruptly rejected the French offer, leaving Napoleon III no choice but to declare war.[56]

This classic example, known in history as the 'Ems Dispatch affair', shows us an unprincipled though brilliant politician. While this case shows how politics controls strategy absolutely, it also demonstrates how a strategist capable of taking responsibility 'blesses' the politician before he takes a move fraught with huge consequences. In short, Ems demonstrates an unrivalled example of cooperation between politics and strategy. There is a huge difference between the 'artistic' falsification of the Ems dispatch and the clumsy fables of 1914 about 'eighty French officers who, in disguise, attempted to cross the German border' or 'several French aircraft that bombed Nuremberg'. This difference represents the difference between Bismarck and Bethmann–Hollweg as politicians, and between Moltke the Elder and Moltke the Younger as strategists.

In the Germany of 1870 (then still Prussia), both politics and strategy were of the highest standard. In the Germany of 1914, however, neither politics nor strategy met any standard.

* * *

Sometimes, a situation occurs when one of the two 'elements of national activity' performs well, while the other lags behind. This lack of synchronisation signifies the frail body of the state organism and the loss of coordination among its parts. It points to a disorder

---

[56] Kersnovski's description of the Ems Dispatch affair is full of incorrect facts (e.g., the meeting took place on 13 July in Berlin and not on 16 in Ems, the telegram was from the king and not the French ambassador, etc.) and details that are not described in Bismarck's memoirs, despite Kersnovski's reference to them.

of the state organism, when its right hand loses the feeling of solidarity with the left one.

The most notorious example of lack of coordination between politics and strategy was offered by Napoleon. The greatest general in history appeared to be a completely incapable politician. He neglected the wise tradition of Richelieu and royal France. The 1813 Battle of the Katzbach and the 1813 Battle of Leipzig were results of his shortsighted politics. Napoleon's foreign policy led to the unification of (instead of fragmentation among) all his enemies. His domestic policy was similarly catastrophic. His civil legislation, based upon the anarchist-individualistic spirit of Rousseau's utopias and an attempt to preserve the Jacobin centralisation of power, destroyed the foundations of the family in France. The hundred thousand French soldiers whom Napoleon led to their deaths in his beautiful (but, in the end, futile) battles are nothing in comparison with the tens of millions unborn because of his legislation. His 'Civil Code' destroyed the French birth rate. We all know the famous words of Lord Castlereagh at the Congress of Vienna: 'Why would we finish France? Let's leave it to its legislation.'

The period of decline in our own state can also be summarised as lack of coordination between politics and strategy. In 1877, our politics performed well (supported by the personal influence of Alexander the Liberator and public patriotism). It had the courage to make a 'great-power-style' decision and declare war on Turkey in defiance of Europe. The strategy, however, was deplorable.

In 1878 strategy improved and the Russian Army was at the walls of Constantinople. But, then, Russia's politics capitulated.

The year of 1905 was marked by complete dissonance when politics simply ignored strategy. It was a mistake to consciously risk a conflict with Japan without completing the Trans-Siberian Railway. It was a mistake to pursue great-power ambitions in the Far East, basing this decision on two or three battalions of the Siberian Army Corps. It was a mistake to take the concessions on the Yalu River[57] without building

---

[57] Concessions on the Yalu River involved forest concessions on the border between China and Korea, obtained in September 1896 from the Korean

STRATEGIYA

an appropriate harbour in Port Arthur. It is a mistake to take a second step without taking a first. Strategy, also, was not at its best, allowing itself to be surprised. Prime Minister Witte[58] and General Kuropatkin deserved each other.

Russian strategy during the Great War was quite mediocre, though it was not as bad as it might seem when judging by the results. Its hands and feet were tied tightly. Russia complied unconditionally with even the most absurd demands of its allies and unquestionably sacrificed its crucial interests for the sake of satisfying their pettifogging, mercantile calculations (under the cover of 'allied common purpose'). We played this pathetic role. We rushed into the fire on the first order from the allies. We were immediately taken in tow by the allies, falling under their complete and absolute influence, enslaving ourselves by the terrible Treaty of London.[59]

This humiliating submission was seen even in little things. For example, Russian generals travelled above the Arctic Circle to the allied conferences in Chantilly, as nobody would even consider organising them in Baranovichi or Mogilev (they could have had a very important role as, in this case, Russia would have been represented by first-class delegates, thus increasing its weight in the negotiations). This small example characterises our inability to maintain Russia's dignity in negotiations with foreigners. Our history is full of conferences in Paris, London, Vienna and Berlin, yet there is not even one 'Peace of St Petersburg' or 'Treaty of Moscow'.[60] Even after a successful war,

---

government by Vladivostok-based merchant Julius Briner for a period of twenty years. The concessions upset Japan (as it saw these as Russian penetration into Korea) and served as one of the causes of the Russo-Japanese War.

[58] Count Sergei Witte was the first Prime Minister of Imperial Russia from 1905 to 1906.

[59] The 1915 Treaty of London.

[60] While Kersnovski tries to make a point, his argument here is not entirely correct. For example, there were the 1762 Treaty of St Petersburg that ended the fighting in the Seven Years' War between Prussia and Russia; the 1875 Treaty of St Petersburg between the Empire of Japan and the Empire of Russia; and the 1881 Treaty of St Petersburg between the Russian Empire and the Qing dynasty.

we went to apologise for our victories to foreign capitals instead of inviting interested foreigners to come to us!

We never knew how to negotiate with foreigners. Hence, we failed to put ourselves in an appropriate position when negotiating during the Great War. We failed to exploit our situation, which in fact was politically favourable. The Allies needed us badly, especially during the first two years of war. We had to *sell* our help to them, just as they sold their help to us.

An excellent example of negotiation was given by Italy, when it stubbornly and shamelessly bargained before entering the war. Political flair has always been honoured by the followers of Machiavelli. From the very beginning, Italy showed its future allies that it would not 'do the donkey work' for them. Thanks to its political flair and will, Italy's weight at inter-Allied conferences was greater than the weight of Russia, even though the contribution of the former was incomparably smaller than the part played by the latter.

On the one hand, the French censored the pre-war plans for our strategic deployment. They defined both the strength of our Northwestern Front and the schedule for increasing its preparedness. As a result, it was the French who predetermined our offensive in Eastern Prussia on the fifteenth day of mobilisation. On the other hand, we were deprived of the right to make any comments or express any requests regarding the famous 'Plan XVII'. For example, one sad consequence of Russia's numerous capitulations to its allies' demands was the Lake Naroch Offensive in March 1916. After all, this disaster was executed on our Western Front to ease the situation of the Allies in Verdun.

Two hundred Russian officers and soldiers hanged themselves on German barbed wire (the Second Army alone lost 140,000 killed and wounded), thereby saving thousands of French lives. Until April 1916, one and a half times more Russians were killed for Verdun than the French. This disastrous offensive, undertaken in the slush of March, had such a significant impact on General Evert[61] that, even in the summer, he categorically refused to renew offensive operations, leaving the small (though victorious) armies of the Southwestern Front unsup-

---

[61] Aleksei Evert was a Russian Imperial general. He commanded an offensive at Lake Naroch in March 1916 together with General Aleksey Kuropatkin.

STRATEGIYA

ported. Without support from the north, they choked on their own victory, and the campaign of 1916 lost all that had been gained.

This was an example of the cruel strategic consequences created by the actions of weak politics, lack of character, and inability to defend one's rights and say 'no' (while explaining the reason for this 'no'). We were unable to say anything to our allies, even when they demanded that we tear pieces of meat out of our living bodies. Moreover, despite our sacrifice, when Marshal Pétain[62] wrote his book 'Verdun' twenty years later, he made no mention at all of those 200,000 Russians who gave their blood and lives on Lake Naroch.

Of all Russian leaders during the Great War, only General Gurko[63] (a strong advocate of Russia's equality of position among the Allies) and Admiral Eberhardt (who commanded the Black Sea Fleet in 1914) were gifted with the necessary political flair and consciousness of statesmanship. {Though it is also important to note the undeniable political talent of General Baratov[64] in Persia.} Immediately after SMS 'Goeben'[65] arrived in the Golden Horn, Admiral Eberhardt realised that the arrival of the German ships would drag the Turkey of Enver Pasha into a war with Russia (and that the main consequence of this war would be the closure of the straits and Russia's complete isolation from the rest of the world). He proposed to attack 'Goeben' in Turkish waters with his five old yet capable ships, intending to prevent Turkey from joining the war by this political preventive measure. While it would disappoint the Sublime Porte and the Young Turks and make Downing Street express its regret at such arbitrary behaviour, Russia

---

[62] Philippe Pétain was a French general during the First World War, known as 'the Lion of Verdun'.

[63] Vasily Gurko was a Russian Imperial general. He served for a brief period as chief of staff of *Stavka* from November 1916 to February 1917.

[64] Nikolai Baratov was a Russian Imperial general who led a very successful campaign in Persia against the Ottoman Empire during the First World War.

[65] SMS *Goeben* and SMS *Breslau* entered the Dardanelles on 10 August 1914, several months before the Ottoman fleet, including German-crewed ships, carried out surprise raids on Theodosia, Novorossiysk, Odessa and Sevastopol, forcing Russia (on 2 November) to declare war on the Ottoman Empire.

would be saved from dying of suffocation. Sazonov,[66] however, did not approve of the operation. In 1878, Russian diplomacy feared the English ironclads; in 1914, it was afraid of its own.

* * *

The failure of politics was also felt during the Civil War.

The tragedy in Russia's case is that the Red side (essentially anti-government) turned to the methods of anarchists, while the anarchistic nature of the White Movement turned out to be the reason for its failure.

In the first year of struggle for the future of Russia, both sides were characterised by anarchism. The Kuban Campaigns were equally improvised by both the Reds and the Whites. However, while the Reds rushed to reject improvisation as soon as possible, and entered the path of organisation, the Whites systematised their improvisation. The bravery of both Kuban Campaigns infused this improvisation with a quality of heroism. Romance replaced politics; voluntarism, regular service; improvisation, statehood.

This was the main reason for the catastrophic outcome of the second year of war, the reason that ruined the 1919 Advance on Moscow. Lack of politics, or its neglect, expressed itself in the lack of order in the occupied territories and the subsequent inability to use their manpower (despite a population of sixty million, the number of bayonets on the front barely reached 22,000). An enormous number of officers was not put to work (in the territory controlled by the Armed Forces of South Russia there were approximately 70,000 officers). The opportunity to create a regular military force and revive the state was missed. The famous proverb 'better late than never' does not apply to politics.

During the Crimean period of the war, anarchism expressed itself in the lack of foreign policy. The real purpose of the 1920 Northern Taurida Operation was similar to the Lake Naroch Offensive, but this time it was aimed at saving Poland.

---

[66] Sergei Sazonov was Foreign Minister of Imperial Russia from November 1910 to July 1916.

STRATEGIYA

Pilsudski[67] was as much an enemy of Russia as Lenin. The fact that Poland involved itself in a war with Soviet Russia was very advantageous for the Armed Forces of South Russia, creating a chance for relief after the winter defeat and the Novorossiysk catastrophe.[68]

The White Movement could exploit the Polish–Soviet war as much as possible in pursuit of its own interests.

Moreover, a Polish defeat could be very advantageous to the White cause. Firstly, it would defeat one of the enemies of the Russian state. Secondly, the conquest of Poland and subsequent advance of the Bolsheviks to the borders of Central Europe (which was shaken by the war and presented nothing less than a huge barrel of gunpowder) would startle France, jeopardising the French-led rearrangement of Europe. It would turn Wrangel[69] in Crimea into the only possible saviour of the situation, allowing him to dictate his terms to the French government.

While a defeat of the Polish would increase the political weight of the Armed Forces of South Russia, their victory would make 'White Russia' obsolete.

General Wrangel failed to realise this and tried to help Poland, basing his decision not on healthy political calculations, but on a romantic (and wrong) assumption that 'anyone who fight against the Bolsheviks is our ally'.

A true politician (who has not only a soul of fire, but also a cool head) in such a situation would not prevent the Red enemy of the Russian state from crushing the Polish enemy of the Russian state—a minus and a minus makes a plus.

---

[67] Jozef Pilsudski was a Polish statesman who served as the head of state (1918–22) and First Marshal of Poland.

[68] The Novorossiysk Catastrophe involved the evacuation of the White Armed Forces of South Russia and refugees from Novorossiysk in March 1920, in which thousands of officers, soldiers and Cossacks of the White Army and civilians were left behind and killed by the Red Army.

[69] After the Novorossiysk Catastrophe, Pyotr Wrangel (who started the First World War in the rank of captain) replaced Lieutenant General Anton Denikin as commander-in-chief of the Armed Forces of South Russia, which changed its name to the Russian Army.

The ideal solution would have been a decision to withdraw the army after the successful operation on 25 May[70] to Crimea, taking with it as many supplies as possible. Safe behind the fortified isthmuses of Crimea, it would be possible to organise the army and wait, indifferent to Polish and French pleas for help (while blindness in politics is pernicious, deafness can sometimes be useful). The answer to the plea from Warsaw would be simple: by imprisoning the forces of General Bredov in concentration camps,[71] Poland deprived itself of the right to be supported by the Russian Army. In a similar vein, Paris should be told that no action to help Poland (and, indirectly, France) would be made before the Russian Army was supplied with the necessary equipment (including sufficient military aircraft to deal with the strong Red cavalry). Such strong language would have an immediate effect and everything required would be supplied without question, allowing the deployment of the whole force (and not a tiny fraction of it) of the Russian Army in the most decisive battle of the liberation war—the Kuban Campaigns.

The help that was offered by France in July–August 1920 was not a 'grant', but 'repayable assistance'—whereby France sought to sell its help for as high a price as possible. At this time, France was in a position where any bargaining was impossible. However, the complete quiescence on the Crimea Front from June to August could have been an excellent weapon to apply political pressure on France. The Crimean government (extremely weak politically) failed to utilise this exceptional political and diplomatic situation during the summer of 1920.

The Polish, however, utilised it well, receiving free help for which, in other circumstances, they would have been asked to pay a very high

---

[70] Kersnovski refers to the first successful stage of the Northern Taurida operation (which eventually became a failure).
[71] Kersnovski refers to the famous 'Bredov March', the retreat following the Novorossiysk Catastrophe, undertaken by 20,000 White soldiers and 7,000 refugees, under the command of General Nikolai Bredov, from south-west Ukraine to Poland in early 1920. However, when the White forces encountered units of the Polish Army, they were disarmed and interned in former German military camps. In August 1920, what remained of Bredov's troops made their way to Crimea to join the Russian Army.

STRATEGIYA

price. The September peace between Poland and Soviet Russia[72] (which presented the Bolsheviks with White Crimea on a silver plate in a noble but unwise gesture) was a lesson that Polish state politics gave to the anti-state politics of the White Movement. The fruits of this anti-state politics were bitter: Wrangel was defeated by Pilsudski, not by Budyonny.[73]

Calling Polish politics 'treacherous' is as unreasonable as complaining about Austrian 'ingratitude' in the Crimean War. The morality of a country cannot be judged by the same rules as the morality of a person—these are two incomparable things.

Chapter 7: Strategy, Operatika[74] and Tactics

Strategy is about waging wars. Operatika is about waging campaigns. Tactics is about fighting battles. Strategy is the competence of the General Staff. Operatika is the competence of the army commander. {The term 'operatika' was introduced to Russian (and therefore international) military science by General Gerua[75] (the author of *Hordes*, Sofia, 1923) and Colonel Evgeny Messner.} Tactics is the competence of everyone down the chain of command—from corps commander to platoon leader.

While the higher end of strategy overlaps with politics, its lower end intersects with operatika. The goal of strategy is to direct operatika

---

[72] Kersnovski refers to the 1921 Peace of Riga, negotiations for which started in Riga in September 1920.

[73] Semyon Budyonny started the First World War in the Russian Imperial Army, eventually joining the Red Army and becoming one of Soviet Russia's military heroes by the end of the Civil War.

[74] 'Operatika' is the original word used by Kersnovski. While the word makes certain sense in Russian and English, implying operational level or operational art, it is not a Russian word. Since Kersnovski did not use it in the sense of 'art' or 'level', it is left here in its original spelling. The absence of the term in Russian military terminology can be explained by the fact that it was introduced by the Russian officers in exile and was never picked up by the Soviet military.

[75] Alexander Gerua was a Russian Imperial general who joined the White Movement and, after the defeat in the Civil War, emigrated to Hungary.

towards the aim defined by politics and win the war by means of successful operations and campaigns.

While the higher end of operatika overlaps with strategy, its lower end intersects with tactics. The goal of operatika is to coordinate tactics and strategy and to achieve victory in the whole operation and the whole campaign by coordinating battles within one meaningful system of time and space.

The goal of tactics is to successfully wage battles (the most elementary military action). To wage a battle successfully, tactics should employ (as well as possible) its two fundamental elements: human (the constant element) and technical means (the variable element).

War is waged not in a vacuum, but on terrain. Therefore, the geographical element (one of the most important and defining characteristics of the leading factor of war—politics) also influences military commanders of different ranks. Strategy must take into account geopolitical conditions; operatika should consider geographical conditions (especially the transportation network); tactics, topography.

While strategy gives operatika its political orientation, operatika in turn provides tactics with its strategic orientation. Since strategy should be subordinated to politics, operatika should be subordinated to strategy, and tactics to operatika.

* * *

In practice, the orderly hierarchy between these three elements of commandership is frequently disturbed. It depends on the character of commandership, which depends on the personality and spiritual quality of the commander.

On the one side, there is commandership based on rationalism, characterised by a tendency to sacrifice strategy for the sake of operatika. On the other side, there is commandership based on intuition that neglects reality, leading to superficial operatika. In the first case we see short-sighted commandership; in the second, far-sighted.

To demonstrate these too extremes of commandership, we will focus on General Ludendorff in the spring of 1918 and General Wrangel during the Civil War. While the former was born to crawl (regardless of his undeniable talents), the latter was born to fly too high.

STRATEGIYA

In preparation for a decisive blow to the Allies, Ludendorff chose to attack the British front in Picardy.[76] His decision shows a clear neglect of the spiritual element (the psychological assessment of the enemy). Ludendorff was a positivist and thus he considered only the material aspects of the situation. He did not take into consideration the character of his enemies and their psychological qualities, otherwise he would have delivered his first and most powerful blow at the French.

Ludendorff failed to take into consideration the traditional British egoism, torpor and mentality of 'it has nothing to do with me',[77] which characterised British commandership throughout the Great War. If the French Army was defeated (at the Chemin des Dames or elsewhere), the Englishmen would retreat to their bases, without thinking twice about helping the French. However, in the opposite scenario, the French would always rush to the help of the English.

Napoleon in 1815 perfectly realised this feature of the British character (he got to know the English well during his campaigns in Spain). Believing in the passivity of Wellington, he struck Blücher first in the Battle of Ligny. His main mistake was that he did not annihilate him, and so, later at Waterloo, 'Marshal Forward'[78] was still able to save the 'Iron Duke'.[79] Wellington, however, as Napoleon rightly judged, never thought of helping the Prussians, either at Ligny or afterwards.

Therefore, Ludendorff's strategy in Picardy was wrong. It repeated the mistake of the 1854 Battle of Inkerman during the Crimean War, just on a much larger scale. Like Menshikov,[80] Ludendorff attacked the

---

[76] Kersnovski refers to the 1918 Operation Michael.

[77] Kersnovski uses the Russian idiom '*moya khata skrayu (nichego ne znayu)*', which describes a 'it has nothing to do with me' mentality. Its literal translation, however, is 'my house is on the outskirts, so I know nothing'. Therefore, when Kersnovski uses it to describe the British, it has two interconnected meanings. The first indirect meaning concerns the British mentality, the second and direct one concerns the geographical position of Britain, which shapes its behaviour.

[78] Nickname of Gebhard Leberecht von Blücher.

[79] Nickname of Arthur Wellesley, 1st Duke of Wellington.

[80] Prince Alexander Menshikov was a Russian Imperial general who was commander-in-chief of the Russian forces in Crimea during the Crimean War.

English first, and like Bosquet,[81] who ran (on foot) to save Raglan[82] on Home Hill, the French corps (this time in automobiles) rushed to rescue Byng[83] and Gough.[84]

Striking his blow at the English, Ludendorff planned to take the path of least resistance: the quality of the English Army was poor in comparison with the French (especially regarding the commanding staff). However, he neglected the irrational element of military affairs in favour of the rational: he neglected strategic considerations (in the wider political sense of this term) in favour of operatika. As a result, his 'path of least resistance' turned out to be the path with the greatest resistance: in Picardy, the Germans had to deal with both enemies, instead of attacking at the Chemin des Dames first and dealing with the French only. {Two months later, Ludendorff decided to attack the French Army. After 1917, the Allies had never been as close to losing the war as at the Chemin des Dames. However, the strength of the Germans then was not the same as it was in Picardy.}

Now we will turn to General Wrangel, whose commandership was completely opposite in character.

In spring 1919, Wrangel advocated the need for an offensive directed at Tsaritsyn[85] to join forces with the army of the Supreme Ruler of Russia, Admiral Kolchak,[86] who was approaching the Volga River. The nature of Wrangel's idea was undeniably 'strategic'.

---

[81] Pierre Bosquet was a French general during the Crimean War. His timely intervention at the Battle of Inkerman in support of the British forces secured victory for the Allies.

[82] FitzRoy Somerset, 1st Baron Raglan, was a British general, commander of the British troops sent to the Crimea.

[83] Julian Byng, 1st Viscount Byng of Vimy, was a British general; in spring 1918 he commanded the Third Army, which was the main target of the German offence during Operation Michael.

[84] Hubert Gough was a British general; in spring 1918 he commanded the Fifth Army, which was the main target of the German offence during Operation Michael.

[85] Stalingrad during the Soviet time, currently Volgograd.

[86] Alexander Kolchak was a Russian Imperial admiral. During the Russian Civil War he established an anti-Communist government in Siberia and was recognised as the 'Supreme Leader and Commander-in-Chief of All Russian Land and Sea Forces' by the other leaders of the White Movement.

However, in this specific case, Wrangel completely neglected operatika and its geographical element, as his plan to join forces with Kolchak was out of time and space.

It was 'out of time', as by the end of April the defeated forces of the Supreme Ruler were withdrawing away from the Armed Forces of South Russia. [...] The siege of Tsaritsyn coincided with Kolchak's retreat from Ufa.

It was also 'out of space', because even in the case of a successful merger with Kolchak [...] the new front line would pass across uninhabited (and waterless) steppes, far away from any political centres of the country. Moreover, this 'deserted' front would not have even a single railway. [...] The only result it could lead to was catastrophe. [...]

Since Ludendorff was looking from the 'bottom up', he missed the perspective of strategy. Since Wrangel was looking from the 'top down', he missed the perspective of operatika. {In analysing a war or operation, we always have to take into consideration the personality, character and spiritual quality of the commander who leads it. The only way to understand it is through the 'prism' of the commander. [...]} [...]

Another important example of the relationship between strategy and operatika that deserves our attention is the Russian commandership during the Great War.

In analysing our plan for strategic deployment in August 1914, it becomes clear that the Russian armed forces had two goals: to defeat the Austro-Hungarian forces and to ease the position of the French forces. The first goal evolved in the context of the Eastern Front of the war and was given to our Southwestern Front. The second goal evolved in the context of the whole Great War and was given to our Northwestern Front.

In 1914, the Russian command acted in 'two dimensions': politico-strategic (the Northwestern Front) and strategic-operational (the Southwestern Front). This situation made the Russian General Staff an 'all-Allies' commander.

The campaign in Eastern Prussia was absolutely necessary. From the strategic perspective, easing the position of France was more important than defeating Austro-Hungary. For the Eastern Front of the Great War (separate from other factors), the Southwestern Front was the main front, and the Northwestern Front was secondary; but for the

Great War itself (in the context of all its theatres) the main role belonged to the Northwestern Front, as it most represented the main principle of war—politics.

When starting an operation, a surgeon should analyse not only the organ he plans to operate on, but also the heart of the patient. The 'organ' of the Eastern Front of the Great War was the Southwestern Front, while its 'heart' was the Northwestern Front.

Let us assume that all forces would be directed towards defeating Austro-Hungary on the Southwestern Front, and the Northwestern front against Prussia would be given a passive role with insufficient forces. Russia would defeat Austro-Hungary, but Germany would defeat France—what would happen next? [...]

From the strategic perspective, our deployment in 1914 perfectly suited the dual nature of the goals defined for the Russian armed forces. From the perspective of operatika, however, the deployment was extremely unsuccessful: the armies were organised according to the same 'template' with unclearly defined main operational lines (on the Northwestern Front there was no line at all, and the line on the Southwestern Front was unclear and wrong). [...]

Moreover, from October 1914, the Russian command was forced to take into consideration a new variable which completely changed the course of war. The entrance of Turkey into the war was the most important political event of the Great War. On the one hand, from this moment Russia was isolated from the rest of the world, doomed to die a slow death by suffocation. On the other hand, the alliance of Turkey with the enemy camp, combined with the extremely favourable diplomatic situation (Britain would be forced to take our side), opened for Russia a window of opportunity to achieve its old great-power ambitions.

Politics and strategy imperiously demanded both a 'surgical operation' to eliminate the looming suffocation from blockade and a turn in the war towards pursuing Russia's ambitions as a great power. What was lost in 1878 began to fall into Russia's hands in 1915, turning the Turkish Front into the main theatre from the perspective of great-power competition. However, while this reduced the importance of the Austro-German Front in the eyes of strategy and politics, from the perspective of operatika this front continued to be the most important one, attracting ninety-five per cent of the total military force.

STRATEGIYA

The political organ of the state (its government) understood (vaguely) the enormous importance of the Turkish Front. In April 1915, it mobilised landing forces (approximately two corps) in Odessa and Sevastopol in preparation for the assault on Constantinople and the crossing of the straits. All Turkish forces were busy fighting in Gallipoli, leaving the Bosporus and Constantinople almost defenceless. Moreover, Russia could count on the possible help of Greece and perhaps Bulgaria.

However, the strategic organ of the state (*Stavka*)[87] was not mature enough to understand the great-power element of politics and the political element of strategy. Confused by the disaster of the Gorlice-Tarnów Offensive,[88] *Stavka* redirected the troops intended for landing in Constantinople (the most important Russian operation in the Great War) to Galicia. In Galicia, however, these two corps produced no effect, [...] simply increasing the number of casualties. At the Bosporus, they could have decided the fate of the Great War; instead, at the San River, they were a grain of sand in the maelstrom of total retreat. [...]

The reason for this blindness was that the commandership of the Grand Duke, General Danilov,[89] like the commandership of General Ludendorff, was rationalistic. They were all good students of Moltke—positivists who a priori deny the importance of the spiritual element and consider only material ones. They simply were not capable of imagining that the fall of Constantinople would bolster public opinion in the country to the extent that nobody would notice the temporary loss of Galicia, the Duchy of Courland and Lithuania. Russia could derive unlimited strength to successfully continue the war. [...]

When Emperor Nicholas II took command over the army, he made the right step in trying to give the war a great-power character. An

---

[87] *Stavka* refers to the general headquarters of the late-19th-century Imperial Russian armed forces.

[88] The Gorlice-Tarnów Offensive was initially conceived as a minor German offensive to relieve Russian pressure on the Austro-Hungarians, but resulted in the Central Powers' chief offensive effort of 1915, causing the total collapse of the Russian lines.

[89] Yuri Danilov was a Russian Imperial general. At the start of the First World War, he was chief of operations (third-in-command) at *Stavka*.

offensive against Constantinople, under the command of Admiral Kolchak, was scheduled for April 1917. God, however, judged differently. All possible deadlines were missed, the country was already suffocating. Torturing strategy does not go unpunished. [...]

* * *

The examples discussed above allow us to judge the nature of the relationship between different elements of commandership.

Politics and strategy, operatika and tactics, are the multipliers of commandership. Each of them contributes a known positive value to the formula of commandership. If one of these multipliers is underestimated and reduced to the level of a 'proper fraction', then the total outcome of the formula—commandership—is reduced. In 1918, Ludendorff underestimated strategy; hence, despite his excellent operatika and tactics, the results were not great—the outcome was less than the values of the multipliers (as always happens with 'proper fractions'). Moreover, if one of the elements of the formula is completely ignored (i.e., regarded as zero), the outcome of the whole formula turns to zero, regardless of the value of the other elements. Wrangel's plan to join Kolchak, discussed above, is the best example of this, as the value of his operatika was zero.

While we find this mathematical metaphor useful, we should warn the reader that it is used for demonstration only. Any attempt to develop it further 'mathematically' would ultimately lead to one of the cardinal sins of military affairs—positivism. There are no more dissimilar concepts than mathematics and military affairs. While mathematics deals with abstract values, military affairs deal with living beings, their virtues and weaknesses. Military values, unlike their mathematical counterparts, do not have common characteristics and are not commensurate. Politics, strategy and tactics are multipliers in one formula, yet they exist in different dimensions and are not commensurate. [...] In mathematics, 'one' is always 'one'; in military affairs 'one' is never 'one'. A political 'one' is not equal to the 'one' of operatika. A spiritual 'one' cannot be comprehended within three-dimensional Euclidean space. [...]

STRATEGIYA

PART THREE: ON WAGING A WAR

Chapter 10: The Principles of Waging a War. *Glazomer*,[90]
Speed, Thrust

Nothing can be taken from or added to the timeless formula of Suvorov. *Glazomer*, speed and thrust were, are and will forever remain the threefold principle of waging both war and battle. These three elements are omnipotent in politics, strategy, operatika and tactics.

* * *

Suvorov puts *glazomer* in the first place. It is both the intention and the assessment of the situation. Speed and thrust are the implementation, i.e., the exploitation of the situation. The wider the element of war, the clearer the primacy of practical judgement. While it is very important in tactics and operatika, in strategy it has sovereign rule. When it comes to politics, it is nothing more than the *glazomer* of the ruler.

Glazomer without speed and thrust leads a battle into stalemate. The 1807 Battle of Eylau is a good example of this situation. The same result was produced by the tardiness of Potemkin, who gave us Ochakov but missed Constantinople.[91] Speed and thrust without *glazomer* produce an irredeemable catastrophe, such as the 1708 invasion of Russia by Charles XII of Sweden or the reckless swoop of Hitler in 1939.

The next element after *glazomer* is speed, which has a special value in operatika.

The final element is thrust, the primary virtue of tactics. In strategy, thrust is sometimes superfluous, as it can impede *glazomer*. In politics, it is frequently pernicious, as it obscures *glazomer* (as was tragically demonstrated by Hitler—a gambler and mystic, who by no means has the personality of a statesman).

Glazomer is a natural virtue, developed by practice. Speed largely depends on technical capabilities (the network of roads and their condi-

---

[90] The Russian word *glazomer* (*coup d'œil*—literally, an ability to measure distances by eye) describes the ability to assess a situation.
[91] Kersnovski refers to the 1787–92 Russo-Turkish War.

tions). When it comes to thrust, this (largely natural) virtue depends directly on the tactics of every given army in a given period. For example, the French Army, which demonstrated an exceptional thrust in the Crimea[92] and Italy,[93] behaved quite passively during the 1870 Franco-Prussian War owing to new regulations introduced two years before the war.

Even a military genius cannot always find harmony between *glazomer*, speed and thrust. On the one hand, Napoleon demonstrated classic examples of harmony in his 1796 Italian Campaign, as well as in 1805 and 1806. On the other hand, in 1813 the same Napoleon demonstrated a complete absence of *glazomer* when he dispersed his forces across German fortresses, entering the Battle of Leipzig from an exceptionally unfavourable position. Similar mistakes can be seen in the conduct of many masters of military affairs (and in most of the cases it is a problem of *glazomer*). Only Suvorov provided an unrivalled model of this harmony during his entire career, from Stołowicze[94] to the famous crossing of the Alps.

Chapter 11: On Coalition War

The main rule of the politician during a coalition war should be *full freedom of* action. States should wage war only when they are required to do so by their interests. They must halt military activities and exit the coalition at the moment when the continuation of war becomes disadvantageous and their interests are not respected by their allies.

States should never sign preliminary agreements or announce solemn declarations about the impossibility of a separate peace. By doing so, we tie our hands (the biggest mistake that can be made by a bad politician) and deprive ourselves of the most precious weapon of diplomatic pressure. We discard the main trump and sign a blank cheque on which unscrupulous allies can write whatever they wish.

When Peter the Great, fighting against Sweden in coalition with England, Denmark, Prussia and Poland, realised that his allies were

---

[92] Kersnovski refers to the 1853–56 Crimean War.
[93] Kersnovski refers to the 1859 Franco-Austrian War.
[94] Kersnovski refers to the 1771 Battle of Stołowicze during the War of the Bar Confederation between the Russian Empire and the Bar Confederation.

playing a double game, trying to benefit from the Russian sacrifices without making an effort themselves,[95] he immediately withdrew from the coalition in 1717 and continued the war by himself.[96] He even started negotiating peace and an alliance with Sweden (which failed owing to the death of Charles XII of Sweden).[97] This great politics suited the great monarch of our great country.

Sazonov enslaved Russia by accepting the Treaty of London in September 1914.[98] He tied Russia's hands and turned the Russian Army into cannon fodder for the benefit of foreign states. It was the worst action ever.

* * *

A strategist, like a politician, must maintain his full freedom of action. He should not tie his hands with preliminary 'military conventions'. These conventions are as unwelcome in strategy as the declarations about the impossibility of a separate peace are unwelcome in politics. Neither numbers, nor timing, nor formal obligations should be promised in advance. While very little should be promised, everything promised should be fully delivered. Every service provided should be immediately followed by an invoice, and every service of the ally should be immediately paid back. If, during a military operation, we are asked to pull chestnuts out of a fire, we have to demand from our allies a pair of fireproof gloves.

Rescuing Verdun in 1916, we sacrificed 200,000 Russian officers and soldiers on German barbed wire at Lake Naroch. We broke our

---

[95] Kersnovski refers to England and Denmark, which constantly switched sides during the 1700–21 Great Northern War.

[96] In fact, Russia never found itself fighting alone in this war. Kersnovski probably refers to the 1717 Convention of Amsterdam, which marked a new alliance between Russia, France and Prussia, leaving out other former allies.

[97] Kersnovski refers to the 1718–19 Congress of Åland, direct peace negotiations between Sweden and Russia.

[98] Kersnovski refers to the 1915 Treaty of London, a secret treaty between the Triple Entente and the kingdom of Italy that brought Italy into the First World War on the Allied side. While it was signed in April 1915, the first proposals were brought to Sazonov in September 1914.

backs for the rest of the campaign, without receiving even a simple thank-you from our allies, let alone compensation. In contrast, when the Allies demanded decisive action from the Italian chief of staff, General Luigi Cadorna, in December 1916, he said that he would not move an inch until they sent him 400 pieces of heavy artillery.[99] His strong language was understood and respected.

The military commander is the complete master of his armed force and his own decisions. He should definitely take into account requests arriving from his allied counterparts and also forward his own requests for their consideration. However, in no circumstances should he endure unasked advice, simultaneously refraining from giving similar advice to others.

The two and a half million Russian soldiers who fell with glory during the Great War demand of us these basic rules of coalition warfare.

PART FOUR: ON MILITARY MAN

Chapter 12: Characteristic of military man

Military virtues can be divided into two categories: characteristics generally required to wear a military uniform honourably in all circumstances, and characterises required to fulfil certain duties in times of war or peace (i.e., general characteristics and specific characteristics).

The three main military virtues are discipline, vocation and sincerity.

While many assume that bravery is the main military virtue, it is, in fact, a derivative of these three main characteristics. It is based within each one of them. A military unit (as well as its men) that preserves discipline under fire is a brave unit (with brave men). A soldier who confidently and ardently believes in his vocation cannot be a coward. Finally, sincerity (the open pursuit of one's religion, views and beliefs) implies openness and straightforwardness and, therefore, it is much higher a quality than bravery [...]. Bravery 'by itself' (so-called 'naked bravery') does not have much value, as it is not connected to any of these three main military virtues. [...]

---

[99] Kersnovski probably refers to the pressure applied by the French and British on Italy in the winter of 1916–17 before the Tenth Battle of the Isonzo.

STRATEGIYA

Chapter 13: Military Ethics and Soldiers' Ethics

We define military ethics as a collection of rules and traditions (codified and uncodified) that should guide adversaries during war. Soldiers' ethics can be defined as rules and traditions observed by all members of a military family in their relations among themselves, as well as in relations between the military and civilians.

* * *

The 'cabinet wars', waged by professional armies in pursuit of state interests from the end of the 17th century and throughout the 18th century, were the golden age of humanity. Wars were waged without hatred of the enemy, as there were no 'enemies'. Instead, there were adversaries, which were determined and ferocious in battle, considerate and courteous after battle, and honourable even in the most dangerous situations.

After the 1799 Battle of Trebbia, Suvorov ordered the return of swords to the Seventeenth Demi-brigade, out of respect for the two-hundred-year-old tradition of glory and valour of the Régiment Royal-Auvergne, from which it was formed. Half a century before that, at Fontenoy, when the British found themselves face to face with the French Guards, who showed no sign of disorder, Lord Charles Hay called to the French commander: 'Order your men to fire!' 'After you, English gentlemen!' replied the French commander, politely saluting with his sword. The subsequent first volley of the British killed hundreds of the French.[100] This 'Après vous, messieurs les Anglais!' has become a common story in the relations between the two nations. A hundred and seventy years later, John French[101] reminded Foch of it when the same British brigade sacrificed itself to cover the French retreat at Ypres.[102]

Contemporary military ethics is just a pale shadow of what was developed by generations of soldiers during the hundred and fifty years

---

[100] Kersnovski refers to the famous story during the 1745 Battle of Fontenoy.
[101] John French, 1st Earl of Ypres was commander-in-chief of the British Expeditionary Force for the first year and a half of the First World War.
[102] Kersnovski probably refers to the 1914 First Battle of Ypres.

of cabinet politics and professional armies. They created enough honour, courage and courtesy to transmit to the hordes of the First Republic (as these hordes were led by the officers and NCOs of the old royal army, who could instil in their subordinates the tradition and spirit in which they were raised).

It was the Revolution of 1789 with its armed 'masses' that inflicted the most severe damage on military ethics. The clashes between the armed French nation and the armed Spanish and Russian peoples resurrected the images of barbaric invasions and religious wars.

Professional (or semi-professional) armies gave wars a touch of humanism which is now completely lost. The 1853–56 Crimean War and the 1859 Franco-Austrian War were the last great wars waged by gentlemen.

The 1870 Franco-Prussian War (and the behaviour of the German armed nation in it) already demonstrated that rules of morality and military ethics are completely incompatible with the mind of the armed masses. [...] By replacing professional 'well-mannered' armies with ferocious people's militias, humanity replaced scourges with scorpions, aggravating the outcomes of war.

At the same time, wars are inevitable (just as death is inevitable), and no paper agreement will help to eradicate them. Therefore, humanity should do as much as it can to ameliorate wars and avoid the gangrene of moral decay—a painful process which can last for many years after the end of a war. No public education can help here. A thousand intellectually developed people, when combined, turn into an ignorant and ferocious mob. The arsonists who torched Leuven,[103] and the executioners who butchered the citizens of Dinant,[104] belonged to the most educated nation in the world. The most decisive factor here is cultivation, which is superior to simple education. Only by discarding the psychosis of the 'armed nation', by restoring to the armed forces their professional character, and by cultivating our lives in the spirit of the Church, will we release ourselves from the noose

---

[103] Kersnovski refers to 1914 Fire of Leuven, when the entire city centre was systematically destroyed by the German occupying forces.

[104] Kersnovski refers to 1914 Battle of Dinant, during which German forces massacred hundreds of Belgian civilians.

thrown round our neck by the doctrinaires of 1789 and their followers. Only when war regains its character of a 'benign ulcer', instead of its current character of a malignant boil, will we be able to speak about military ethics again.

* * *

Soldiers' ethics are a collection of rules (most of which are unwritten) that are used by all members of a military family in their relations among themselves.

Full members of the military family (so-called 'adult members') are soldiers by vocation (the officer corps, regular enlisted soldiers and volunteers). Only they can be asked to comply with the demands of soldiers' ethics in all their strictness.

The interaction of junior staff with the seniors (subordinates with superiors) is generally defined by the military code—the 'written' rules of soldiers' ethics. However, the approach of the seniors to the junior staff is a less clear area.

Every commander, regardless of his position (including the chief of staff), should always remember that he is not simply 'commanding' his troops, but that the commandership is an honour and privilege. He must remember this in peacetime (respecting the military dignity of his subordinates), as well as (and even more so) in times of war, when the honour of the company, battalion or army entrusted to him is inextricably linked to his own honour and his good name in the eyes of future generations. [...]

Chapter 14: Mind and Will

All virtues of a military man, whether major or secondary, are rooted in two main foundations: 'mind' and 'will'. A balance between these two, so perfectly represented in Peter I, Rumyantsev and Suvorov, provides an ideal type of military man, an ideal type of leader.

Yet, usually one of these elements outweighs another, thus creating either a predominantly 'mind-led' commander (Bennigsen)[105] or a

---

[105] Levin August von Bennigsen was a German general in the service of the Russian Empire from 1773 to 1818.

predominantly 'will-led' commander (Blücher). While the former are planners, the latter are executors.

Sometimes we come across the complete hypertrophy of one element at the expense of another. This can lead to either a purely 'mind-led' commander characterised by atrophy of will (Kuropatkin, Alekseev)[106] or a purely 'will-led' commander characterised by atrophy of mind (Charles XII of Sweden). Both cases have a pathological character and inevitably lead to catastrophe.

While mind without will is an absolute zero, will without mind is a negative value.

In the sphere of commandership, a predominance of will over mind gives better results than a predominance of mind over will. An energetically implemented mediocre decision will always offer better results that an unimplemented or hesitantly implemented ideal decision. A constantly circulating copper coin is more useful than chervonets buried in the ground.[107] Schwarzenberg[108] had a much higher intellect and better scientific background than Blücher, yet the fiery soul and indomitable will of 'Marshal Forward' raised his commandership (despite Brienne[109] and Montmirail)[110] to a level much higher than that of Schwarzenberg. Mackensen, who had no higher military education, proved himself much better than the academic and erudite General Klyuev.[111]

A will-led orientation (originating in the heart and therefore irrational) characterises leaders of military art. A mind-led orientation, how-

---

[106] Mikhail Alekseev was an Imperial Russian Army general during the First World War and the Russian Civil War. Between 1915 and 1917 he served as Tsar Nicholas II's chief of staff of the *Stavka*.
[107] *Chervonets* is the traditional Russian name for large foreign and domestic gold coins.
[108] Karl Philipp, Prince of Schwarzenberg was an Austrian general, who was commander-in-chief of the allied Grand Army of Bohemia and the senior of the allied generals who conducted the campaign of 1813–14.
[109] The 1814 Battle of Brienne.
[110] The 1814 Battle of Montmitail.
[111] Nikolai Klyuev was an Imperial Russian Army general during the First World War and the Russian Civil War. He is best known for surrendering his army to the German forces during the First World War.

STRATEGIYA

ever, is intellect-based and therefore characterises leaders of military science. Will manifests itself less frequently than mind, as it is more difficult to develop. While mind can be developed by education, will must be cultivated.

A will-led orientation is an inherent characteristic of the Russian people, who created a world power in conditions in which any other nation would fail. Our history presents titans of will, such as Alexander Nevsky, Patriarchs Hermogenes[112] and Nikon,[113] and Peter the Great. It is also the inherent characteristic of Russian commandership.

Saltykov[114] successfully defended his army from the encroachments of Daun[115] and St Petersburg's 'conference'.[116] Rumyantsev successfully completed the apparently hopeless siege of Kolberg,[117] even though the military council called by him proposed (on three occasions!) to lift the siege. Suvorov demonstrated inhuman power of will at Izmail,[118] and superhuman will in the crossing of the Alps. Who can really appreciate the will of Barclay,[119] who went against the tide to save Russia against its

---

[112] Patriarch Hermogenes was the Patriarch of Moscow and all Russia from 1606 to 1612. It was he who inspired the popular uprising that put an end to the Time of Troubles.

[113] Patriarch Nikon was the seventh Patriarch of Moscow and all Rus' of the Russian Orthodox Church from 1652 to 1666. He was renowned for the introduction of many reforms which eventually led to a lasting schism known as Raskol in the Russian Orthodox Church.

[114] Count Pyotr Saltykov was a Russian statesman and a military officer. From 1759, he was commander-in-chief of the Russian Imperial Army during the Seven Years' War.

[115] Count Leopold Joseph von Daun was an Austrian field marshal of the Imperial Army in the Seven Years' War. He was accused by his Russian allies of seeking to win the war by sacrificing Russian blood.

[116] 'Conference' was a term used to describe the military council organised and led by Elizabeth of Russia during the Seven Years' War.

[117] The 1759–61 Siege of Kolberg during the Seven Years' War.

[118] The 1790 Siege of Izmail during the 1787–92 Russo-Turkish war.

[119] Prince Michael Andreas Barclay de Tolly was a Baltic German field marshal and Minister of War of the Russian Empire. In 1812, Barclay de Tolly was commander of the First Army of the West, the largest army to face Napoleon. He initiated a scorched earth policy from the beginning of the campaign, though this made him unpopular among Russians.

own will? By sacrificing Moscow, Kutuzov demonstrated a more powerful will than Napoleon at the Battle of Leipzig. [...]

The structure given to our army by Alexander I of Russia, after the end of the Napoleonic Wars (a period that is incorrectly called 'Arakcheyevism'),[120] did not contribute to the creation and, more importantly, the promotion of powerful characters. As a result of this, Paskevich[121] froze the army, Milyutin instilled into it a corrupting 'non-combatant spirit', Vannovsky[122] depersonalised it, and Kuropatkin demoralised it. This impoverishment of the military spirit was only one of the facets of the general spiritual impoverishment of our nation, the general destitution of the Russian state.

Will-led characters were obviously present in both the Crimean War (Kornilov,[123] Nakhimov,[124] Muravyov,[125] Bebutov)[126] and the 1877–78 Russo-Turkish War (Radetsky,[127] Gurko,[128] Skobelev,

---

[120] Kersnovski refers to Count Alexey Arakcheyev, a Russian general and statesman. As a very powerful figure in the court of Alexander I, he conducted several important military reforms.

[121] Count Ivan Paskevich was a Russian Imperial military leader, whose career spanned the period from the 1805 Battle of Austerlitz (when he was a young lieutenant) to 1854, when he was the commander of the Army of the Danube, which was then engaging the Turks in the initial stages of the Crimean War.

[122] Pyotr Vannovsky was Minister of War of the Russian Empire from 1881 to 1898.

[123] Vladimir Kornilov was a Russian vice admiral. During the Crimean War, he was responsible for the defence of Sevastopol.

[124] Pavel Nakhimov was one of the most famous admirals in Russian naval history, best remembered as the commander of naval and land forces during the Siege of Sevastopol during the Crimean War.

[125] Nikolay Muravyov-Karsky was a Russian Imperial general known for his successful command during the 1855 Siege of Kars during the Crimean War.

[126] Vasiliy Bebutov was a Russian Imperial general known for his victory in the 1854 Battle of Kurekdere during the Crimean War.

[127] Fyodor Radetsky was a Russian Imperial general known for his command during the third and fourth stages of the Battle of Shipka Pass during the 1877–78 Russo-Turkish War.

[128] Count Iosif Romeyko-Gurko was a Russian Imperial general known for his victories during the 1877–78 Russo-Turkish War.

STRATEGIYA

Ter-Gukasov).[129] However, weakness of will already started to emerge at the Danube and Crimea (thanks to the completely impersonal Gorchakov),[130] and in 1877, when Grand Duke Nicholas Nikolaevich the Elder[131] and Loris-Melikov[132] almost lost the war. In the 1904–5 Russo-Japanese War, fussy and weak-willed Kuropatkin clipped the wings of strong-willed Gripenberg.[133] And finally, in the Great War, will-less Alekseev undid the brilliant successes of the 1916 campaigns by his hesitations, persuasions, negotiations and conversations.

There were will-led characters in the Great War: Lechitsky,[134] Plehwe,[135] Yudenich, Brusilov,[136] Count Keller.[137] However, Russian commandership in this war was defined by a lesser type of commander: Alekseev, Ruzsky, Evert. As a result of the general destitution of the Russian state, Alekseev in the *Stavka* was matched by Belyaev[138] as

---

[129] Arshak Ter-Gukasov was a Russian lieutenant general. He was the Yerevan Forces' commander of Russia's Army during the 1877–78 Russo-Turkish War.

[130] Prince Alexander Gorchakov was a Russian diplomat. He served as Foreign Minister of the Russian Empire from 1856 to 1882.

[131] Grand Duke Nicholas Nikolaevich of Russia (the Elder) was the third son of Tsar Nicholas I of Russia. He commanded the Russian Army of the Danube in the 1877–78 Russo-Turkish War.

[132] Count Mikhail Loris-Melikov was a Russian Imperial statesman and general. In the 1877–78 Russo-Turkish War, he commanded a separate corps d'armée on the Turkish frontier in Asia Minor.

[133] Oskar Ferdinand Gripenberg was a Russian Imperial general of Finnish-Swedish origin. He commanded the Russian Second Manchurian Army during the 1904–5 Russo-Japanese War.

[134] Platon Lechitsky was a Russian Imperial general. He was famous for a series of successful operations during the First World War.

[135] Paul von Plehwe was a Russian Imperial general of Baltic German origin. He commanded the Fifth Army, which he led in the Battle of Galicia and the defence of Łodz during the First World War.

[136] Aleksey Brusilov was a Russian Imperial general most noted for the development of new offensive tactics used in the 1916 Brusilov Offensive.

[137] Count Fyodor Keller was a Russian Imperial general. He was famous for a series of successful operations during the First World War.

[138] Mikhail Belyaev was a Russian Imperial general and statesman. He was

Minister of War, Khabalov[139] as the commander of the Petrograd military district was matched by Protopopov[140] as Minister of the Interior.

* * *

The superiority of the 'will-led' type over the 'mind-led' type is especially apparent in a comparison between Russian and German commanders in 1914.

Our commanders were lacking in faith in their own profession and the great future of their motherland and army. They lacked the will to engage with the enemy and win—win no matter what. Neither hot nor cold (used to receiving ranks, distinctions and higher positions easily and effortlessly), they did not feel the honour and glory of the military profession. They forgot that they did not simply 'command' their troops, but that the commandership was their honour and privilege—and this honour must be paid for. [...]

The German commanders of 1914 recall our commanders of the great century. When at Stallupönen[141] General François[142] received an order to retreat, he replied: 'Report that General von François will withdraw when he has defeated the Russians!' This recalls pretty well the reply of Kamensky[143] at Oravais:[144] 'Fellows, we will not retreat until we smash the Swedes!' While Kamensky won at Oravais, François retreated without defeating the Russians. [...]

The Germans drew their spiritual strength from their national doctrine—'Deutschland über alles' (Scharnhorst, Moltke and Schlieffen

---

chief of staff of the Imperial Russian Army from 1914 to 1916 and the last Imperial Minister of War from 3 January 1917 to 28 February 1917.

[139] Sergey Khabalov was a Russian general of Ossetian origin and the commander of the Petrograd military district in 1917.

[140] Alexander Protopopov was a Russian publicist and politician who served as Minister of the Interior from September 1916 to February 1917.

[141] The 1914 Battle of Stallupönen.

[142] Hermann Karl Bruno von François was a German general during the First World War and is best known for his key role in several German victories on the Eastern Front in 1914.

[143] Count Nikolay Kamensky was a Russian Imperial general best known for his performance during the 1808–9 Finnish War.

[144] The 1808 Battle of Oravais during the Finnish War.

STRATEGIYA

merely symbolised it, while Fichte, Clausewitz and Treitschke[145] were its true inspirers). They were like Dokhturov,[146] Kamensky and Miloradovich,[147] who drew their strength from Suvorov's 'We are Russians, God is with us!' [...]

The problem of will is, first and foremost, a problem of military ethics and the cultivation and organisation of the officer corps. [...]

PART FIVE: ON ARMED FORCES

[...]

Chapter 18: Army in the State

The organisation of the armed forces is necessary to address two important questions. The first concerns the organisation of the permanent part of the army (by offering it the status of a military caste or a military estate of the realm). The second concerns the impermanent part of the army—the armed nation.

Any army consists of generals, officers and soldiers. Officers shape the soldiers, but also produce generals. Therefore, officers are the foundation of the armed forces.

The officer corps can have the character of a closed caste, a serving estate of the realm, a professional union, a party organisation, or a group of people connected by individual contracts of service with the state. This last option represents the situation of the 20th century, reducing armed forces to a pile of grains of sand, a house of cards ready to fall at the first gust of wind.

Organising armed forces in the form of a party organisation is similarly impossible. Party ideology is incompatible with military spirit and ethics (the deplorable history of the Red Army serves as a good example of that). Moreover, party ideology is incompatible with Christian

---

[145] Heinrich von Treitschke (1834–96) was a German historian, political writer and National Liberal member of the Reichstag during the time of the German Empire.

[146] Dmitry Dokhturov was a Russian Imperial general and a prominent military leader during the 1812 Patriotic War.

[147] Count Mikhail Miloradovich was a Russian Imperial general prominent during the Napoleonic Wars.

Orthodoxy; hence, it is incompatible with Russianness itself. We should reject this suicidal and deeply amoral possibility.

A union of professionals also contradicts the very nature of the Russian armed forces, the only army in the world that has never employed landsknechts or condottieri mercenaries.

Therefore, the only options available to Russia are either a military caste or an estate of the realm.

The German Army of 1914 was a caste. On the one hand, a son of a *feldwebel* (NCO), General Alekseev, the chief of staff of *Stavka* during the Great War, would have failed to attain even the rank of junior officer in the German Army before the war. On the other hand, raised in the spirit of the caste, German generals would never betray their Kaiser (though it did not stop their Kaiser from betraying them, when he disgracefully abandoned his defeated army).

The strength of a military caste lies in its highly developed sense of military ethics and caste solidarity (all for one and one for all). Caste is a monolith, infused by a unifying spirit, where the senior members have unquestionable authority in the eyes of the junior members, and the junior members enjoy the comradely respect of the senior members.

The other side of a caste is its closedness and alienation from the people. The traditions of the German Army go back to the Livonian Brothers of the Sword, which ruled harshly from their castles over the enslaved tribes of the Baltics and Prussia. The caste's honour and arrogance deeply contradict the Russian Orthodox worldview, and we (whose ancestors blunted their swords with Teutonic skulls) should not adopt the customs of the descendants of those who were defeated on the Neva River,[148] and in the battles of the Ice,[149] Wesenberg[150] and Grunwald.[151]

In organising the family of the Russian military estate of the realm, we should take the model of shaping the officer corps [...] developed by Peter I—a model that proved itself successful for more than a century. This system allows for the creation of the required monolith without establishing conditions unnatural to the Russian mind and its understanding of castes.

---

[148] Kersnovski refers to the 1240 Battle of Neva.
[149] Kersnovski refers to the 1242 Battle on the Ice.
[150] Kersnovski refers to the 1268 Battle of Wesenberg.
[151] Kersnovski refers to the 1410 Battle of Grunwald.

STRATEGIYA

The main foundation of the officer corps should be the tightly bound and cohesive family of regimental officers. Special attention should be paid to the creation of regimental officer societies and recognition of the great significance of regimental commanders. It is important to remember that while the company is an administrative unit and the battalion is a tactical unit, the regiment is a spiritual unit. Military units above the level of regiment (brigade, division and corps) also have a purely tactical and operational character.

The spirit of an army is created in its regiments, just as the spirit of the navy is created in its ships. A regimental commander should be the full and sole master of his unit, complying with the orders of brigade, division and corps commanders only in terms of hierarchy. He should have the rights of a ship's captain on the open sea. A commander shapes his regiment according to his own image, engraving his own (sometimes indelible) imprint. He has an enormous influence on his troops, an influence that he would not usually have in subsequent higher positions of authority. The memory of an extraordinary commander will be passed down in a regiment from generation to generation, outliving any memory of a similarly extraordinary commander of a division or corps. [...]

* * *

The armed forces are one of the elements of politics (similar to education, administration and diplomacy). The statement that 'an army is apolitical' is ridiculous. The army is the very expression of armed politics.

Some describe the army as 'a sword of the state'. This is completely wrong. While the sword is a blind weapon (an insensible tool that can be controlled by anyone, including a criminal), the army is a living organism.

The army is not a sword. It is an arm holding the sword. It is a living arm, directed by the will of the head. And the Tsar is the head.

Taking into consideration the conditions of Russia, this is the only possible formula. Any other is simply excluded. Communist barbarism, democratic senility, totalitarian godlessness: they all equally corrode the state and corrupt the army. We should reject these products of soullessness and stupidity.

ANTON KERSNOVSKI

The triune covenant 'for Faith, Tsar and Fatherland'[152] is not an empty utterance. Its magnificent simplicity is rooted in the thousand-year-old existence of Russia, during which our *Tsars*, taking strength from the *Christian Orthodox faith*, created for us the greatest country in the world—our *Fatherland*.

This is our politics. Once we understand it, we will become invincible. [...]

PART SIX: ON DOCTRINE

Chapter 22: On National Doctrine and Military Doctrine

When speaking about 'doctrine', we imply a collection of views on a specific question, as well as a collection of 'methods' to address this question.

Military doctrine is the worldview of a given nation regarding its military situation, and it is one of many facets of this nation's national doctrine. Hence, the main characteristic of a military doctrine is its nationality. Military doctrine is a derivative of the historical, cultural and military traditions of a given nation, its political, geographical and ethnic conditions, and the spirit and psychology of the people (or peoples) that created it. In sum, military doctrine mirrors the spiritual face of a nation.

Any importation or cultivation of foreign doctrines, therefore, violates the spirit of the nation (and this violation never leads to anything good). It is possible to cut a small piece (military doctrine) from the monolithic block of a foreign national doctrine and solder, stick or tie this piece to our national monolith (our national doctrine). However, this connection will never be strong, even if we prepare and shape our monolith in advance to better absorb this foreign piece. This connection will always remain mechanical and constantly fail to become an organic part of our monolith and its conditions. Therefore, the resulting artificial, bastard doctrine will fail to withstand any serious challenge.

In analysing military doctrine, we refer to two distinctive types: the German doctrine and the Russian one. While the former is character-

---

[152] Kersnovski refers to the famous motto used in the Russian Imperial armed forces in the 19th century: *Za veru, Tsarya i otechestvo*.

ised by its strict system and technical attention to detail, the latter is the doctrine that was bequeathed to us by Suvorov.

* * *

German national doctrine is a deification of the Germanism (Deutschtum) that subjugates all other peoples (especially Slavic peoples) whom German culture sees as manure. The German nation sees itself as a nation of rulers (Herrschervolk)—a noble northern race that alone should rule the enslaved world. [...]

This leads to the main characteristics of German military doctrine: the 'strategy of annihilation' and the 'battle of annihilation' (Vernichtungsschlacht). German 'integrated warfare' and its associated cruelty and terror are intended to paralyse the civilian population (the lower race of the enemy state), killing its will to resist and bringing the war to a quick end.

When a German officer cuts the fingers of Belgian girl, his action is completely justified by Clausewitz and Ludendorff with their inhumane theory of 'integrated warfare', by Treitschke with his theory of Herrschervolk, and Fichte with his 'Addresses to the German Nation'. In German eyes, children's blood is a quite permissible method of war, as long as these children belong to a lower (not German) race.

These misanthropic tendencies influence not only the ethical side of German military doctrine, but also its very nature. The main subconscious driving force of a German strategist or tactician is his self-image as a Spartiate armed with a scourge, setting out to chastise a crowd of trembling helot slaves. [...] This image would never enter the mind of a Russian, French or British soldier.

In their attempt to transfer the tactics of François[153] and Morgen[154] (the famous 'methods' of the German manuals and doctrines) onto 'Russian soil', Russian rationalists and positivists take military doctrine

---

[153] Hermann Karl Bruno von François, the German general. He is best known for his key role in several German victories on the Eastern Front in 1914.

[154] Curt von Morgen was a German general during the First World War. He is best known for his key role in several German victories on the Eastern Front in 1914.

out of the context of the national doctrine that gave it birth. They try to take François and Morgen without Clausewitz, Treitschke, Fichte, the Livonian Brothers of the Sword [...] who created them. They are trying to take the leaves without the tree and its deep roots. They try to stick a branch from a foreign tree into Russian soil, naively hoping that it will take root.

A German considers himself 'superhuman' (failing to notice that by doing so he reduces himself to the level of 'subhuman'). A Russian, however, following the timeless definition of Dostoyevsky (the brightest spokesman of Russian national doctrine), considers himself 'all-human'.[155] While German national doctrine was forged by Thor, Russian doctrine was inspired by Christ. While the Sword of Sigurd was forged by the dwarf Mime, the Sword Kladenets of Ilya Muromets was blessed by a saint. These two different origins have shaped two very different military doctrines.

This is why Russian military doctrine (Russian national military doctrine) must have the same imprint of humanity that, during the last eleven centuries, has made Russia 'Jehovah's favoured warrior'.

Chapter 23: Russian National Military Doctrine

The core facet of the Russian national monolith, which is called Russian national military doctrine, is *the superiority of spirit over matter*.

This superiority of eternity over mortality was felt by Russian cannoneers at Zorndorf,[156] when they kissed their cannons goodbye, refusing to retreat while their bodies were cut to pieces by the sabres of Seydlitz's[157] cuirassiers (a situation that would force any German to retreat or surrender). The same feeling drove Rumyantsev at the Kagul

---

[155] All-human (*Vsechelovek*). Kersnovski refers to Dostoyevsky's speech 'Pushkin (A Sketch)' where he writes: 'To become a true Russian, to become entirely Russian, may, after all, mean nothing more than to become the brother of all men—an all-human, if you like.'

[156] Kersnovski refers to the 1758 Battle of Zorndorf during the Seven Years' War, which was fought between Russian troops commanded by Count William Fermor and a Prussian army commanded by Frederick the Great.

[157] Friedrich Wilhelm von Seydlitz was a Prussian officer. He commanded the cavalry charge against Russian troops during the 1758 Battle of Zorndorf.

STRATEGIYA

River[158] to confront 200,000 Turks with his force of 17,000. It inspired Suvorov's pen when he wrote the timeless texts of 'The Science of Victory' as well as his sword, when he brightened the path for his bogatyri[159] in the grey morning of Rymnik,[160] during the hot days at the Trebbia River,[16] and in the gloom of Alpine nights.[162] This superiority of spirit over matter burnt like a fire in the souls of the musketeers of Miloradovich and the infantrymen of Dokhturov,[163] the riflemen of Yudenich and the assault troops of Kornilov.[164]

The foundations of Russian national military doctrine were, are and will be as follows. As Christian Orthodox people, we see war as an evil (an immoral disease of humanity)—an immoral legacy of our ancestors' sins. There are no pompous words, no paper treaties or other attempts to bury our head in the sand that will help us to prevent this evil. Like a dragon drawn on the door that will not save a Chinese household from plague, the parchment of the 1928 Paris agreement (the Kellogg-Briand Pact) will not save humanity from war. And if this is so, then we need to prepare ourselves to face this evil and strengthen the state organism, increasing its resistance (the job of lawmakers and politicians).

Military art and military science (when the role of the latter is to serve the former) have a national character and are based on the spiritual characteristics and qualities of every given nation. A Russian

---

[158] Kersnovski refers to the 1770 Battle of Kagul during the 1768–74 Russo-Turkish War.

[159] A *bogatyr* (plural *bogatyri*) is a stock character in medieval East Slavic legends, akin to a Western European knight errant.

[160] Kersnovski refers to the 1789 Battle of Rymnik during the 1787–92 Russo-Turkish War.

[161] Kersnovski refers to the 1799 Battle of Trebbia between the joint Russian and Habsburg army under Alexander Suvorov and the Republican French army of Jacques MacDonald.

[162] Kersnovski refers to the famous Suvorov manoeuvre through the Swiss Alps in 1799.

[163] Dmitry Dokhturov was a Russian general and a prominent military leader during the Patriotic War of 1812.

[164] Lavr Kornilov was a Russian general in the Imperial Russian Army during the First World War and the leader of the anti-Bolshevik Volunteer Army during the Civil War.

Moltke simply cannot exist, just as a German Suvorov is impossible. The German who moved furthest towards 'Suvorovism' (as far as his German roots allowed him) was Blücher. The Russian who moved furthest towards 'Moltkeism' (as far as his Russian roots allowed him) was Milyutin.[165] Our true teachers are Peter I, Rumyantsev, Suvorov and a few other Russian generals and political figures, as well as several foreigners who are not organically alien to us. *Foch and Moltke cannot be our true teachers. The most they can do for us is to be our tutors.*

Suvorov's 'The Science of Victory' is incomparably greater and clearer than the sophistry of Clausewitz, the scholasticism of Schlieffen, and even the brilliant metaphysics of Seeckt. Russian national military doctrine always puts the principle of quality and selectivity at the core of military *organisation*. For example, Peter I built his army on the principle of selectivity by attracting Russian nobility to the service. The Russian Army of the 18th century was, first and foremost, selective—something that explains well its great victories during this century.

The very organisation of Russian armed forces (either in the Tsardom of Muscovy or in the Russian Empire) has always been a consequence of *our identity*. 'Our similarities with the people of Europe are weak,' wrote Rumyantsev in his 'Thoughts on the Organisation of Armed Forces'. The decline of our army began when we started to imitate foreign patterns during the rule of Paul.

In *waging our wars*, we must try to avoid its inhumane forms. We need to reject with disgust the 'Clausewitzism-Leninism' theory of integrated warfare with its terrorisation of the enemy civilian population. We need to remind ourselves of the words of our second (after Lazarev)[166] commander-in-chief in the Caucasus, Tsitsianov:[167] 'The rule of the Russian soldiers is to beat their adversaries when required,

---

[165] Count Dmitry Milyutin was Minister of War and the last field marshal of Imperial Russia. He was responsible for sweeping military reforms that changed the face of the Russian Army in the 1860s and 1870s.

[166] Ivan Lazarev was a Russian Imperial general. He was famous for his pivotal role in the Russian Empire's annexation of the Kingdom of Kartli and Kakheti (eastern Georgia) in the last years of the 18th century.

[167] Prince Pavel Tsitsianov was a Georgian nobleman and a prominent general of the Imperial Russian Army. From 1802 to 1806 he served as the Russian commander-in-chief in the Caucasus.

but not to destroy them, as the Russians do not know how, after beating an enemy, not to annex its land to the Russian state, and, as such, everyone has to preserve his property.'

If we read these golden words with our spiritual, and not bodily, eyes, we will realise their timeless meaning. We do not need to annex an enemy's lands (as long as these lands are not our property previously stolen from us); the spiritual annexation of foreigners to our culture is sufficient. And this can be possible only in the absence of mutual resentment and unhealed wounds.

While striving not to be inhuman in foreign lands, we should also avoid being animals in relations with our own motherland. We must wage our wars in a way that aggravates and depletes the organism of our state as little as possible. This can be achieved only when as many professionals as possible are kept in their original places, whether these are farmers or railwaymen, craftsmen or traders. The fatal mistake in 1916 of being mesmerised by armed hordes should be avoided at all cost.

If we want our army to give everything it can, then it should be used accordingly. And Russian national military doctrines provide several rules for that. The first is synthesis—to see 'the affair as a whole' (the closest alternative to this rule in foreign doctrines is Verdy's[168] 'de quoi s'agit il?', implemented by Foch). The second rule includes '*glazomer, speed and thrust*'. Finally, there is victory, which '*must be achieved with the minimum of blood*'.

* * *

The educative side of our doctrine has always promoted our religious roots and national pride. '*We are Russians, God is with us!*'—this was the teaching of Suvorov. This is why his science turned into the real 'Science of Victory'. Its every word was easily accepted by the artless hearts of Russian bogatyri, turning Suvorov's whole teaching (one of the purest creations of Russian genius) into the greatest monument of Christian Orthodox Russian culture. Military schools should open special faculties to teach Suvorov's 'The Science of Victory'. This

---

[168] Julius von Verdy du Vernois was a German general, known for both his military writings and his service on Helmuth von Moltke the Elder's staff during the Franco-Prussian War.

teaching, however, should neither fanatically follow its literal interpretation nor heretically 'correct it to fit contemporary conditions'. Only in that way can it be learned by the officers, allowing them to transfer it to their soldiers.

Furthermore, our doctrine is characterised by the need for a mindful approach to affairs (*'every soldier understands his manoeuvre'*) and lower-rank initiative (either due to Peter's requirement *'not to stick to the manual as a blind person sticks to a wall'* or Suvorov's advice that *'a local commander is in the best position of judgement ... if I say the left, but it should be the right—do not listen to me'*). The approach to higher-rank initiative should be that of Rumyantsev's *'do not get into the details of possible scenarios which any reasonable commander can deal with by himself, and do not tie his hands'*.

From this and all other perspectives, the only way (the only remedy) is a return to the way pointed out to us by Peter, Rumyantsev and Suvorov, the way from which we were diverted a hundred and fifty years ago by the spontoons of the Gatchina Guards.[169]

---

[169] Kersnovski refers to the Gatchina Guards, which were sponsored by Paul I of Russia who favoured the Prussian line infantry drill.

5

THE SCIENCE OF WAR

ON SOCIOLOGICAL RESEARCH INTO WAR[1]

Nikolai Nikolayevich Golovin

CHAPTER ONE

THE POSSIBILITY AND NECESSITY OF THE SCIENCE OF WAR (THE SOCIOLOGY OF WAR)

1. The Science of Waging War (the Theory of Military Art) and the Science of War

The greatest military thinker of the first part of the 19th century, Clausewitz, writes: 'Positive theory is impossible. With materials of this kind we can only say to ourselves that it is a sheer impossibility to construct for the art of war a theory which, like scaffolding, shall

---

[1] This chapter is based on the translation of selected chapters from Nikolai Golovin, *Nauka o voyne: O sotsiologicheskom izuchenii voyny* [The science of war: On sociological research into war], (Paris: Izdatel'stvo Gazety 'Signal', 1938).

ensure to the chief actor an external support on all sides. In all those cases in which he is thrown upon his talent he would find himself away from this scaffolding of theory, and in opposition to it, and, however many-sided it might be framed, the same result would ensue ... talent and genius act beyond the law, and theory is in opposition to reality.' {Clausewitz, 'On War', translated by Voyde.}[2]

Indeed, military science, which restricts the area of its research solely to the analysis of the ways to wage wars, has only ever been able to ascertain that there is no universal way to victory. Even today, this type of science can only reaffirm the conclusion, so vividly formulated by Napoleon, that 'the situation determines a war'. At the end of the 19th century, General Dragomirov,[3] one of the greatest authorities in military affairs, wrote in his remarkable critique of Leo Tolstoy's 'War and Peace': 'nowadays, no one would even think of arguing that there could be a military science'. {Dragomirov, 'War and Peace', p. 48.}[4]

Yet, despite the conclusion made by the great military scholars, such as Clausewitz and Dragomirov, regarding the possibility of a positive military science, this verdict should be reconsidered.

'War is a mere continuation of policy by other means,' writes Clausewitz. And these are the thoughts of John Stuart Mill on politics as a science, which can be equally extended to military science: 'The condition indeed of politics, as a branch of knowledge, was until very lately, and has scarcely even yet ceased to be, that which Bacon animadverted upon, as the natural state of the sciences while their cultivation is abandoned to practitioners; not being carried on as a branch of speculative inquiry, but only with a view to the exigencies of daily

---

[2] Golovin refers to the first Russian translation (from German) by General Carl Voyde of Clausewitz's *On War* published in 1902. The quote comes from Carl von Clausewitz, *On War*, vol. 1, translated by Colonel J. J. Graham (London: Kegan Paul, Trench, Trübner and Co., 1908), p. 106.

[3] Mikhail Dragomirov was a Russian Imperial general and military theoretician. In addition to many significant original works, he was the first to translate Clausewitz's *On War* into Russian (from French) in 1888.

[4] Mikhail Dragomirov, *Razbor romana 'Voyna i Mir'* [An analysis of the novel 'War and Peace'], (Kiev: Izdanie Knigoprodovtza N. Ya. Ogloblina, 1895), p. 46.

practice, and the *fructifera experimenta*, therefore, being aimed at, almost to the exclusion of the *lucifera*. Such was medical science, before physiology and natural history began to be cultivated as branches of general knowledge. The only questions examined were, what diet is wholesome, or what medicine will cure some given disease; without any previous systematic inquiry into the laws of nutrition, and of the healthy and morbid action of the different organs, on which laws the effect of any diet or medicine must evidently depend ... No wonder that when the phenomena of society have so rarely been contemplated in the point of view characteristic of science, the philosophy of society should have made little progress; should contain few general propositions sufficiently precise and certain, for common inquirers to recognize in them a scientific character. The vulgar notion accordingly is, that all pretension to lay down general truths on politics and society is quackery; that no universality and no certainty are attainable in such matters. What partly excuses this common notion is, that it is really not without foundation in one particular sense. A large proportion of those who have laid claim to the character of philosophic politicians, have attempted, not to ascertain universal sequences, but to frame universal precepts. They have had some one form of government, or system of laws, to fit all cases; a pretension well meriting the ridicule with which it is treated by practitioners, and wholly unsupported by the analogy of the art to which, from the nature of its subject, that of politics must be the most nearly allied. No one now supposes it possible that one remedy can cure all diseases, or even the same disease in all constitutions and habits of body ...'[5]

These words of Mill can unreservedly be considered as a description of military science before Clausewitz, and can also be applied, with minor reservations, to our times. Until today, there has been no clear understanding that a positive military science is possible, so long as its aim is restricted to the analysis of war as a process of human social life,

---

[5] Golovin refers to the Russian translation (by V.N. Ivanovskii) of Mill's *A System of Logic*, published in 1900. The quote comes from John Stuart Mill, *A System of Logic, Ratiocinative and Inductive: Being a Connected View of the Principles of Evidence and the Methods of Scientific Investigation* (London: Longmans, Green, and Co., 1895), pp. 571–72.

leaving analysis of the ways to wage wars to the theory of military art. This distinction between the nature of two different aims requires a division of contemporary military art into two fields: 'the science of waging wars', which is a theory of military art; and 'the science of war', which is one of the positive social sciences.

If we accept this division as our starting point, then Clausewitz's denial becomes irrelevant.

In fact, nowadays, the consistent patterns of social life are a generally accepted fact. The distinctive proof of these patterns is given to us by statistics. Is there enough ground to suggest that the phenomenon of war represents any exception in this matter? The obvious answer is no. [...]

2. The Science of War Should Be a Sociological Analysis of War

The reason that many military scholars deny the science of war is hidden in an incorrect assumption that science must give military art some sort of immutable rules. In his 'A System of Logic', Mill states: 'It is not necessary to even the perfection of a science, that the corresponding art should possess universal, or even general rules. The phenomena of society might not only be completely dependent upon known causes, but the mode of action of all those causes might be reducible to laws of considerable simplicity, and yet no two cases might admit of being treated in precisely the same manner. So great might be the variety of circumstances on which the results in different cases depend, that art might not have a single general precept to give, except that of watching the circumstances of the case, and adapting our measures to the effects which, according to the principles of the science, result from those circumstances. But although, in so complicated a class of subjects, it is absurd to lay down practical maxims of universal application, it does not follow that the phenomena do not conform to universal laws.' {Mill 'A System of Logic'.}[6]

The extent to which Mill is right can be supported by the example of Clausewitz. Placing at the core of his work the idea that war must be studied as a phenomenon of social life {'war is part of the intercourse of the human race'}, he defined the aim of his study as 'to

---

[6] The quote is from Mill, *A System of Logic*, p. 572.

explore the nature of military phenomena to show their affinity with the nature of the things of which they are composed'.

Having so wide and truly scientific an aim, the treatise of Clausewitz, who denied, as we saw above, the very possibility of a positive military science, represents the first attempt to create this science in the shape of a 'science of war'. It is very significant that Clausewitz found no other title for his treatise but 'On War'. This title is missing only one word: 'science'. Clausewitz, who denied the very possibility of a positive military science, humbly assumed that his treatise was just a 'review'. The fact that Clausewitz's 'On War' is, still today, the only systematic work on the 'science of war' gives this book a very unusual longevity for a military-scientific treatise. While many brilliant works devoted to the analysis of the ways of waging war quickly become outdated, Clausewitz's 'On War' has continued to attract the attention of wider and wider circles. The war of 1914–18, like the war of 1870–71, pushed military science towards a more detailed examination of Clausewitz.

Therefore, Clausewitz should be considered as the father of the positive 'science of war', and any researcher who desires to analyse war not through a narrow 'utilitarian-military' prism, but through a 'purely scientific' one, must study his work in detail. Such a researcher, like Clausewitz, must understand that the main goal of the analysis of war is its examination as a phenomenon of social life. In other words, a pure science of war should be a **sociological** analysis with the aim of examining the processes and phenomena of war from the point of view of their existence, coexistence, similarity and consistency. In doing so, there is no choice but to follow Clausewitz, who, as we saw above, defined the goal of his classic treatise as 'to explore the nature of military phenomena, to show their affinity with the nature of the things of which they are composed'.

Commenting on the quotation above from Mill's work about the utilitarian character of any science in its initial stages of development, I said that with minor reservations his description can be applied to contemporary military science. I would like to clarify my conclusion. This comment was intended to emphasise the idea that right until today the majority of researchers into war have been analysing the 'ways' to wage wars, and not wars themselves. Yet, when the most outstanding of these researchers go deeper in their analyses, they transcend this framework, penetrating, as Clausewitz did, the field of the 'pure' science of war. Therefore, the 'science of war', even if not distinguished as a separate field of knowledge, already partially exists, scattered throughout the science of waging wars.

The collection of material for the science of war has been occurring in all fields of military science. I do not have enough space to recall all the works whose authors tried to find a way towards positive knowledge. Yet, I will name the most outstanding ones: 'The First Attempts of Military Statistics' by a professor of the Russian Military Academy, Dmitry Milyutin (who was awarded the title of count and was one of the closest assistants of Alexander II of Russia);[7] 'Études sur le combat' [Studies of war] by a French military writer, Colonel Ardant du Picq;[8] 'Geschichte der Kriegskunst im Rahmen der politischen Geschichte' [History of warfare in the framework of political history] by a professor of the University of Berlin, Hans Delbrück;[9] 'Témoins: Essai d'analyse et de critique des souvenirs de combattants édités en français de 1915 à 1928' [Witnesses: Analytical and critical essays on soldiers' recollections] by a professor of Williams College in Massachusetts, Jean Norton Cru.[10]

Among all these different attempts to find a way towards positive knowledge about war, special attention should be given to the treatise of a Russia military scholar, General Genrikh Leer, who lived half a century after Clausewitz. He writes: 'Strategy, in its narrow interpretation, is a treatise on operations in the theatres of military action ... Strategy, in its wider interpretation, is a synthesis of all military affairs, its generalisation, its philosophy. Like philosophy, which aspires to

---

[7] Dmitry Milyutin, *Pervyye opyty voyennoy statistiki* [The first attempts of military statistics], 2 vols. (St Petersburg: Tipografiya Voyenno-Uchebnykh Zavedeniy, 1847–48).

[8] Ardant du Picq, *Études sur le combat: Combat antique et combat moderne* (Paris: Hachette and Dumaine, 1880).

[9] Hans Delbrück, *Geschichte der Kriegskunst im Rahmen der politischen Geschichte* (Berlin: George Stielke Verlag, 1922–27).

[10] Jean Norton Cru, *Essai d'analyse et de critique des souvenirs de combattants édités en français de 1915 à 1928*, (Paris: Les Étincelles, 1929).

explain the phenomena of the world, strategy, which is (in its widest interpretation) a philosophy of military affairs, has the aim of explaining military phenomena (not only each one of them, but also, and more importantly, the general connection between them).'[11]

In this wider definition of strategy as an aggregation of all other military sciences, it is impossible to miss the desire of Leer's great mind to create a military science that would be considered as one among other positive sciences. Yet, even with this wide definition of the aim of strategy, it is probably more a philosophy of military art, rather than a science of war, as this is how Leer's outstanding course on strategy can best be characterised.

In his later works, Leer went further. In his books 'The Methods of Military Sciences'[12] and 'Core Issues'[13] he already started to investigate the 'pure' science of war.

The aim of the 'pure' science of war, as we stated above, should be the examination of war as a phenomenon of social life, and not only an analysis of ways to wage wars. This is why the term 'strategy' does not fully express its nature. Its new and better name was coined at the beginning of this century by another Russian military scholar, General Nikolai Mikhnevich, during his lecture on the need to create a 'sociology of war' at the Russian Military Academy.

As the phenomena of war follow certain patterns, the science of war (the sociology of war) should aspire to discover **laws**. Meanwhile, the science of waging wars (the theory of military art), even with its broadest generalisations, can only lead to **principles**. And the distinction between these two is significant.

A law represents a constant, definitive and immutable relation between the phenomena of nature and human life, which exists by virtue of constant and immutable relations between the forces and factors that create these phenomena.

---

[11] Genrikh Leer, *Zapiski Strategii* [Notes on strategy], 3rd edn (St Petersburg: Obschestvennaya Pol'za, 1877), p. 1.

[12] Genrikh Leer, *Metod voyennykh nauk (strategii, taktiki i voyennoy istorii)* [The methods of military sciences (strategy, tactics and military history)], (St Petersburg: Tipografiya S.N. Khudekova, 1894).

[13] Genrikh Leer, *Korennyye voprosy (voyennyye etyudy)* [Core issues (military studies)], (St Petersburg: Tipografiya V. Bezobrazova i Komp., 1897).

A law only confirms the fact of existence, coexistence, consistency or similarity between phenomena, without defining any aims of activity. Being independent of our will (for law, will is only part of the phenomenon), law is absolute.

A principle represents a mere generalisation. While any principle concerns only the nature of the phenomenon, it is directly connected to the setting of the aim. It constitutes the main idea that has to be followed in the process of solving known questions in military affairs. It constitutes a regulator of creativity, which by no means restricts this creativity. It includes, according to Leer, 'the aim that has to be achieved'.

Consequently, a principle, unlike law, does not confirm any fact, but proposes something, even if conditionally. 'A proposition of which the predicate is expressed by the words **ought** or **should** is generally different from one which is expressed by **is** or **will be**.' {Mill, 'A System of Logic'.}[14]

4. The Sociology of War Is Necessary for the Theory of Military Science Itself

The theory of military science, i.e., the science of waging wars, can find its true starting point only in the science of war, which analyses the nature and essence of the phenomenon of war. The science of war should offer a rationale for those cases where the theory of military art derives its conclusions. The sociology of war will have no lesser significance in cases where the science of waging wars draws its conclusions directly from experience (inductively). In these cases the ideal outcome will be a complete interconnection of these inductive conclusions with the conclusions deduced from the statements made about war by science. However, even in cases where this ideal is impossible, the sociology of war, based on analysis of the nature of the phenomena of war, will offer significant benefits **by pointing every generalisation towards the framework within which it holds true**. And this by itself is a very significant goal.

On closer examination, the theory of military art (the science of waging wars) turns into systems of military art, which can be under-

---

[14] The quote comes from Mill, *A System of Logic*, pp. 619–20.

stood as more specific theories of military art, i.e., theories, conclusions and teachings, which correspond to a given stage of development of the art. For example, in military history we can identify the five-march system[15] of waging war, the line of battle system, the system of cohort tactics, the system of strike or fire tactics, etc.

The existence of a general theory of military art, as well as its forms (the systems of military art), is a completely valid phenomenon and is a strictly logical consequence of the science of war. In all fields of knowledge, in addition to pure science, there are applied sciences with more specific generalisations and, finally, purely practical generalisations, applicable only within narrow frameworks. The main condition for the application of the latter is not to use them outside the frameworks within which they are applicable, i.e., they need to be used scientifically.

Denial of the theory and its systems can be explained by the fact that military history attests to the existence of multiple cases in which an unskilful application of theory or some system was the main reason for defeat. Hence, fear of similar consequences suggested the idea of the theory's or system's failure. What happened is a misconception common to the human mind: a misuse of theory and systems, their occasional falsity, or routine use {Military history shows that the damage caused by systems is mainly the outcome of either their falsity or routine use.}, i.e., the application of an outdated system, despite the appearance of new factors—all these give rise to a sceptical attitude towards the very existence of theory and systems, and towards their useful and, more importantly, entirely natural right to exist. **The incorrect use of weapons has led to an incorrect conclusion about their usefulness**.

By analysing the nature of the phenomena of war, the sociology of war would provide an opportunity to define the frameworks within which the application of a given theory or system is scientific.

This science not only would offer a rationale for the theory of military art, but it would also define the boundaries

---

[15] The 'five-march system' is a military logistics concept that implies that an army should not proceed further from its supply depots than five marches at most. If it is necessary to proceed further, a fresh line of supply depots should be formed.

within which the application of theoretical generalisations is valid, i.e., scientific.

In this way, the sociology of war would provide the required combination, justifying and organising its conclusions and generalisations. Such combination is absolutely necessary for any field of knowledge. According to Auguste Comte, there are very few people whose minds can fully embrace one science (or field of knowledge), even if this field of knowledge is only part of a larger whole. The majority of people limit themselves to studying one more-or-less large part, without troubling themselves with how their specific work fits into the general system. It is necessary to get rid of this evil before it grows successfully on a large scale. Otherwise, the human mind can end up losing itself in a labyrinth of specific analyses.

The history of the development of science shows that every time when the human mind concentrates its attention on form, and not on essence, a new scholastic field appears.

This phenomenon is especially distinctive of the philosophy of the Middle Ages, preoccupied as it was by dialectic tendencies and methods. Since the essence of ideas was defined by dogma and thinking was free only in the field of methods of explanation and application, people discussed, argued and made endless conclusions without considering the principles that stand above any analysis. All human activity during the Middles Ages developed in this direction. This misuse of dialectics led to many different nuances of analysis, multiple divisions and subdivisions, the general transformation of logical reasoning into a linguistic contrivance, and an extreme misuse of the forms of thinking at the expense of thought itself; or, in one word, it led to formalism. {See Fouillée, 'Histoire de la philosophie'.}[16]

All this characterises the so-called period of 'classic scholasticism', i.e., a period when scholasticism powerfully ruled all fields of knowledge. Obviously, this is not characteristic of the contemporary state of military knowledge. Yet, it is impossible to miss the presence of a 'scholastic feature' in many works of military literature. The reason for that is the same reason behind the Middle Age's scholasticism.

---

[16] Golovin refers to Alfred Fouillée, *Istoriya filosofii* [History of philosophy], translated by P. Nikolaev (Moscow: Tipolitografia V. Rikhter, 1893).

A science that defines the analysis of the nature of war as its aim would help military knowledge to finally leave the period of scholasticism. By conducting a scientific evaluation of the principles of military art, this science would allow military thought to come out of the field of artificial generalisations, subdivisions, etc., and advance along the way of robust and strictly scientific analysis.

Any science consists of two factors: (1) systematically arranged information; and (2) the complex of methods and techniques used to collect this information (the methodology and logic of a given science). Mill writes as follows: 'A science may undoubtedly be brought to a certain, not inconsiderable, stage of advancement, without the application of any other logic to it than what all persons, who are said to have a sound understanding, acquire empirically in the course of their studies. Mankind judged of evidence, and often correctly, before logic was a science, or they never could have made it one. And they executed great mechanical works before they understood the laws of mechanics. But there are limits both to what mechanicians can do without principles of mechanics, and to what thinkers can do without principles of logic. A few individuals, by extraordinary genius, or by the accidental acquisition of a good set of intellectual habits, may work without principles in the same way, or nearly the same way, in which they would have worked if they had been in possession of principles. But the bulk of mankind require either to understand the theory of what they are doing, or to have rules laid down for them by those who have understood the theory. In the progress of science from its easiest to its more difficult problems, each great step in advance has usually had either as its precursor, or as its accompaniment and necessary condition, a corresponding improvement in the notions and principles of logic received among the most advanced thinkers. And if several of the more difficult sciences are still in so defective a state; if not only so little is proved, but disputation has not terminated even about the little which seemed to be so; the reason perhaps is, that men's logical notions have not yet acquired the degree of extension, or of accuracy, requisite for the estimation of the evidence proper to those particular departments of knowledge.' {Mill, 'A System of Logic'.}[17]

---

[17] The quote comes from Mill, *A System of Logic*, p. 6.

This statement of Mill is relevant when applied to the logic of every field of knowledge.

The character of scientific techniques (methods) depends a great deal on the specific features of the subject analysed; hence, every science has its own methodology. This leads to the conclusion that a comprehensive study of the methods (methodology) of a science, i.e., a study of their features, created under the influence of the subjects analysed by this science, is **impossible without knowledge of the subject of analysis**. {See Vvedensky, 'Logika'.}[18] This implies that the very development of the theory of military art (the science of waging wars), which already depends closely on the development of its methodology, also **depends on science, which has the aim of analysing the nature of the phenomena of war, i.e., the sociology of war**. [...]

6. The Boundaries of What Is Possible in the Development of the Sociology of War

Having acknowledged the possibility and necessity of the existence of the sociology of war, we now need to clarify the boundaries within which its existence is possible; as the complexity of the subject analysed would definitely touch on them. A clear understanding of what this science can give would immediately permit the rejection of any excessive and precocious demands.

Any science passes through three stages of development:

1) Collection of information,
2) Systematisation, development of methods, and analysis,
3) Prediction.

It is obvious that a science reaches its final stage only when it is fully developed. The extent to which a science can reach the final stage differs among various sciences. The more complex and numerous the elements of the phenomenon analysed are, the more input

---

[18] Golovin refers to Alexander Vvedensky, *Logika kak chast' teorii posnaniya* [Logic as part of the theory of understanding], (Petrograd: Tipografia M.M. Stasyulevich, 1917).

data are required to create an accurate prediction. The science of war, like other social sciences, deals with infinitely complicated phenomena, whether in regard to the number of relevant elements or their complex qualities. As a result, the development of all social sciences lags significantly behind in comparison with other fields of human knowledge. In this respect, they are in the last place among sciences. Consequently, it becomes clear that the science of war is not yet ready to answer all questions posed it. Moreover, the complexity of the analysed phenomenon forces us to recognise an additional limitation. A science about society (and about war) would never achieve the same level of development and precision that simpler sciences, e.g. mathematics, have achieved. The phenomena are so complex that their analysis into all relevant elements would never be possible. And such analysis is absolutely necessary to isolate the unknown variable which we are seeking to predict. Owing to the complexity of the phenomena, the unknown factor would remain also in other variables, creating a quadratic equation, to which mathematics, as a simpler science, can offer a solution. The science of war, however, even at the peak of its development, would achieve its third stage of prediction only to some extent, and will always remain predominantly in the field of analysis. Mill writes: 'There is, indeed, no hope that these laws, though our knowledge of them were as certain and as complete as it is in astronomy, would enable us to predict the history of society, like that of the celestial appearances, for thousands of years to come. But the difference of certainty is not in the laws themselves, it is in the data to which these laws are to be applied. In astronomy the causes influencing the result are few, and change little, and that little according to known laws; we can ascertain what they are now, and thence determine what they will be at any epoch of a distant future. The data, therefore, in astronomy are as certain as the laws themselves. The circumstances, on the contrary, which influence the condition and progress of society are innumerable, and perpetually changing; and though they all change in obedience to causes, and therefore to laws, the multitude of the causes is so great as to defy our limited powers of calculation. Not to say that the impossibility of applying precise numbers to facts of such a description would set an impassable limit to the possibility of calculating them beforehand, even if the powers of the human intellect were otherwise adequate to the task.' {Mill, 'A System of Logic'.}[19]

The boundaries within which the development of the science of war can occur cannot be wider than the boundaries defined for sociology by the Italian sociologist Vilfredo Pareto. {Vilfredo Pareto, 'The Mind and Society', New York: Harcourt, Brace and Co., 1935.}[20] According to him, sociology is 'logical-experimental' and can be based only on the observation of present events or on the description of past ones. It cannot allow any speculative generalisations, moral teachings or anything else that transcends the boundaries of the analysis of real facts. In other words, sociology cannot be built on the basis of principles or general ideas, deduced a priori.

According to Pareto, sociology, limited by this framework, merely ascertains the facts of similarities or mutual dependencies. Owing to the extreme complexity of the phenomena of social life and the extremely large number and variability of its elements, sociology will never be able to deduce generalisable conclusions which are absolutely valid in all specific cases. The conclusions of sociology will always be characterised by a certain degree of uncertainty.

However, acknowledgement of these boundaries for the sociology of war I propose does not contradict the legitimacy and necessity of its creation. The idea of Pareto quoted above about 'ascertainment of the facts of similarities or mutual dependencies' can be reformulated in simpler words as 'ascertainment of the existence and coexistence of phenomena and consistencies and similarities among them'. Since these constitute the aim of any science, the limits suggested by Pareto should merely be taken as the boundaries of what is possible in the development of the sociology of war, and not as a suggestion that it is unnecessary. [...]

7. The Creation of the Sociology of War Requires a Colossal Collective Work

The extreme complexity of the phenomena of war demands an especially extensive preparation of material necessary for the creation of

---

[19] The quote comes from Mill, *A System of Logic*, p. 572.
[20] Vilfredo Pareto, *The Mind and Society*, vol. IV (New York: Harcourt, Brace and Co., 1935).

the sociology of war. Lack of understanding of the need for this field of science (so prevalent in the wider scientific community) has led to a detrimental situation in which there is no scientific institution that has gathered all the literature relevant to this subject. For example, the National Library of France does not have a full collection of the classic works of Hans Delbrück.

If this is the situation with one of the classic bodies of work on the sociology of war (and there are not so many of them), then the situation with regard to less significant works is even more detrimental. [...]

I can prove that out of ten published works that should be foundational for the sociological analysis of war, at least three would be missing in such world-leading cultural centres as the British Library or the Library of Congress. The same situation would be found by a researcher wishing to borrow these ten books from the library of any institution of military science.

It is important to note that the sociology of war, more than any other science, needs to be based on the works of all nationalities. Only a truly international framework would set it free from the myths and distortions that inevitably accompany any national approach to the study of war. This is why I believe that the first step towards the creation of the sociology of war should be the establishment of a special library (at least at one scientific institution in the world) that would include absolutely all works relevant to the development of the science I propose. Such a library would include a very large number of volumes, because, as I have already stated above, the information required for sociological research is dispersed across multiple treatises on military art (the science of waging war).

The dispersion of this information suggests the first necessary stage of work. The information, spread among different treatises on the science of waging war, military and general history, sociology and other sciences that study different fields of social life, must be assembled and evaluated in terms of its value to the sociology of war. [...]

The classification of information required by the sociology of war is especially important, as any new science should use all sources available in all languages. [...]

In an attempt to draw the attention of the scientific community to the necessity of such colossal preparatory work, without which war, as

STRATEGIYA

a social phenomenon, would remain unstudied, I presented a special paper at the 12th International Congress on Sociology that took place in August 1935 in Brussels. {The paper was later published in the military-scientific journal 'Osvedomitel', no. 4, pp. 1–13, published by the Russian Military-Scientific Institute in Belgrade in 1937.} Pointing out that such work cannot be done by the private initiative of a few individuals, I insisted that it could be carried out only by the efforts of an organised scientific collective. I finished my paper with the following words: 'All this colossal work requires significant efforts and resources. It requires full independence that would free it not only from chauvinism, but also from nationalism. This is why the success of the sociological study of war as I have suggested can only be secured by the establishment of a special international institute.

'I completely understand the difficulties attached to the establishment of this special scientific institute for the sociological study of war. However, I believe that, sooner or later, humanity will get there. Meanwhile, I will allow myself to express a humbler request—the establishment of a department of the "sociology of war".

'Only the establishment of this department, at least at one university that has a faculty of social sciences, can be considered as the first true attempt to scientifically study war as a phenomenon of human social life. I have no doubts that such an initiative would be quickly imitated and war would stop being "la grande inconnue" [the great unknown].'

As could be expected, my presentation was a lone voice in the wilderness.

[…].

CHAPTER THREE

IMPLEMENTATION OF PSYCHOLOGICAL ANALYSIS

1. The Element of Danger Characterises the Ambience Within Which the Activity of a Soldier Takes Place

The importance of the 'spirit of the forces' has always been highly appreciated by all military commanders over the ages and in all nations.

All great commanders tried, first and foremost, to increase the spiritual power of their soldiers. Suvorov's 'The Science of Victory' is all about different methods intended to psychologically empower soldiers. We will not tire the reader with various quotations from authoritative military thinkers, as they would never end. Acknowledgement of the important role that the 'spiritual element' plays in achieving victory has become a truism in the theory of military art.

Every page of military history supports the argument that the psyche of the forces plays the most important role in all phenomena of war.

Even though the need for psychological analysis of war has been acknowledged by all military practitioners and scholars, very little scientific analysis has been done in this field. While there is a large body of literature analysing the 'external' side of war, there are very few treatises that focus on the analysis of its spiritual side. Indicative of this lack is the fact that, even today, no military academy has a faculty of military psychology. The only attempt to conduct a complete psychoanalysis of battle was made by a French major, Ardant du Picq. {Ardant du Picq, 'Études sur le combat'.[21] An attempt to continue the work of Ardant du Picq was made by the author of this work. In 1907, the Imperial Military Academy published his work titled 'The Analysis of Battle: The Analysis of the Activity and Characteristics of Man, as a Soldier'.}[22]

Nevertheless, human activity during war occurs in a very specific ambience. Clausewitz writes: 'The combat begets the element of danger, in which all the activities of war must live and move, like the bird in the air or fish in the water. But the influences of danger all pass into the feelings, either directly—that is, instinctively—or through the medium of the understanding. The effect in the first case will be a desire to escape from the danger, and, if that cannot be done, fright and anxiety. If this effect does not take place, then it is courage, which is a counterpoise to that instinct. Courage is, however, by no means an act of the understanding, but likewise a feeling, like fear; the latter looks to the physical preservation, courage to the moral preservation.' {Clausewitz, 'On War'.}[23]

---

[21] Ardant du Picq, *Études sur le combat*.
[22] Nikolai Golovin, *Issledovaniye boya: Issledovaniye deyatel'nosti i svoystv cheloveka kak boytsa* [The analysis of battle: The analysis of the activity and characteristics of man as a soldier], (St Petersburg: Ekonomicheskaya Tipo-Litografiya, 1907).
[23] The quote comes from Carl von Clausewitz, *On War*, p. 103.

STRATEGIYA

The great influence which fear has on human activity in battle can be seen in the observation that 'fear attacks and nullifies every effort of the will in such a manner that it has always been esteemed a deed of heroism to combat and subdue it utterly'. {Mosso, 'Fear', translated by Rezolyion-Soshalskaya.}[24]

'Essentially, soldiers are always afraid,' writes Abbé Bessières in his memoirs about his time serving in the 1914–18 War as a medic in the French Army; 'only civilians, journalists, members of parliament, senators, and ministers who wander around the front line are not afraid and do not run away.' {Bessières, 'Le Chemin des Dames'.}[25]

Even the most courageous people must take this nature of warfare into consideration. Skobelev,[26] who was beloved by his men, used to say: 'There is no one who does not fear death; if someone tells you that he is not afraid, spit in his eye, as he lies. I, too, fear death no less than others. However, there are those who have enough power of will not to show it, and there are those who cannot stand it and run away out of fear of death. I have the power of will not to show that I am afraid; however, the internal struggle is powerful and it constantly affects the heart.' {*Historical Bulletin*, January 1895, 'Etudes from the Past' by D. Obolenskii.}[27]

Calling a spade a spade, Du Picq states: 'Man has a horror of death. In the bravest, a great sense of duty, which they alone are capable of understanding and living up to, is paramount. But the mass always cower at the sight of the phantom, death.' {Ardant du Picq, 'Battle Studies: Ancient and Modern Battle', translated by Puzyrevskii.}[28]

---

[24] Golovin refers to the Russian translation (by A.K. Rezolyion-Soshalskaya) of Mosso's *Fear*, published in 1887. The quote comes from Angelo Mosso, *Fear*, translated by E. Lough and F. Kiesow (London: Longmans, Green, and Co., 1896), p. 275.

[25] Golovin referes to Albert Bessières, *Le Chemin des Dames: Carnet d'un territorial* (Paris: Bloud and Gay, 1918).

[26] Mikhail Skobelev was a Russian Imperial general famous for his heroism during the 1877–78 Russo-Turkish War.

[27] Golovin refers to Mikhail Skobelev quoted in Dmitry Obolenskii, 'Zapiski iz proshlogo' [Studies from the past], *Istorishekii Vestnik* [Historical Bulletin], vol. LIX, January–February–March 1895, p. 105.

[28] Golovin refers to the Russian translation (by A.K. Puzyrevskii) of Ardant

Fear is one of the strongest emotions. Chronologically, in any living being, this emotion appears first. {Ribot, 'The Psychology of Emotions'.}[29] 'The emotion of terror originates in the apprehension of coming evil. Its characteristics are a particular form of pain or misery; the prostration of the active energies, and the excessive hold of the related ideas on the mind.' {Bain, 'Psychology'.}[30] 'If we apply the test of the submergence of pleasure, we shall reckon it [terror] one of the most formidable visitations of human suffering.' {Bain, 'Psychology'.}[31]

The physiological accompaniments of fear create an extremely depressive condition. We will not enter into a detailed discussion about the emotion of fear, and will refer those interested in this topic to works that focus on psychology. {Mosso, 'Fear'; Bain, 'Psychology'; James, 'Psychology';[32] Ribot, 'The Psychology of Emotions'.} Instead, we will suggest the main conclusion of these works. It can be formulated as follows: the emotion of fear serves to cloud the mind and paralyse the conscious will of a person; as a consequence, the motives of a person who is under the influence of fear are transferred into the area of the unconscious. [...]

---

du Picq's *Études sur le combat: Combat antique et combat moderne*, published in 1893. The quote comes from Ardant du Picq, *Battle Studies: Ancient and Modern Battle*, translated by John N. Greely and Robert C. Cotton (New York: Macmillan Company, 1921), p. 94.

[29] Golovin refers to Théodule-Armand Ribot, *Psikhologiya tchuvstv* [The psychology of emotions], translated by M. Goldsmit (St Petersburg: Tipografiya A.A. Porokhovschikova, 1898).

[30] Golovin refers to the Russian translation (by A.M-v) of Alexander Bain's *Mental and Moral Science*, which was published under the title 'Psikhlogia' [Psychology] in 1881. The quote is from Alexander Bain, *Mental and Moral Science: A Compendium of Psychology and Ethics* (London: Longmans, Green, and Co., 1868), p. 232.

[31] The quote is from Bain, *Mental and Moral Science*, p. 234.

[32] Golovin refers to the Russian translation (by I. Lapshin) of William James's *Psychology: Briefer Course*, which was published under the title 'Psikhlogia' [Psychology] in 1905.

2. The Drama Experienced by a Soldier is a Struggle between Two Desires: 'To Win' and 'to Evade' the Danger

'Man does not enter battle to fight, but for victory. He does everything that he can to avoid the first and obtain the second.' {Ardant du Picq, 'Battle Studies'.}[33]

The courage of a savage is undoubtable. The constant struggle with the danger presented by nature, animals and humans hardens the savage's fortitude and teaches him to value life. Fear constantly pursues him; he is in constant danger; he is always alert. He can trust no one, and no one can trust him. {John Lubbock, 'The Origin of Civilisation'.}[34] As a consequence, the savage must get used to danger and develop a habit of fearlessness.

While it may seem that given the savage's unlimited courage, an open battle would be a common thing for him, the reality is different.

According to Tylor, 'the savage or barbarian is apt to fall on his enemy unawares, seeking to kill him like a wild beast'. {Tylor, 'Anthropology'.}[35] [...]

A war between barbarians (similar to what frequently happens also in our time) is a war of ambushes between groups of people, in which each of them chooses to attack and kill not an adversary, but a victim. Du Picq explains: 'Because the arms are similar on both sides, the only way of giving the advantage to one side is by surprise. A man surprised needs an instant to collect his thoughts and defend himself; during this instant he is killed if he does not run away. The surprised adversary does not defend himself; he tries to flee. Face to face or body to body combat with primitive arms, ax or dagger, so terrible among enemies without

---

[33] The quote is from Ardant du Picq, *Battle Studies*, p. 43.

[34] Golovin refers to John Lubbock, *Doistoricheskie vremena ili pervobytnaya epokha chelovechestva* [The origin of civilisation and the primitive condition of man], translated by D. Anuchin (Moscow: Tipografiya A.I. Mamontova, 1876).

[35] Golovin refers to the Russian translation (by I. Ivin) of Edward Burnett Tylor's *Anthropology: An Introduction to the Study of Man and Civilisation*, published in 1898. The quote is from Edward Burnett Tylor, *Anthropology: An Introduction to the Study of Man and Civilisation* (London: Macmillan and Co., 1881), p. 223.

defensive arms, is very rare. It can take place only between enemies mutually surprised and without a chance of safety for anyone, except in victory. And still ... in case of mutual surprise, there is another chance of safety; that of falling back, of flight on the part of one or the other; and that chance is often seized.' {Ardant du Picq, 'Battle Studies'.}[36]

The desire of soldiers to evade danger, alleviating the struggle as much as possible, is proved by the whole history of the development of weapons. The whole meaning of this history can be expressed in a few words: **man contrives to kill his enemy, while avoiding being killed**. He comes with a cudgel against a stake, with arrows against the cudgel, with a shield against the arrows, with a shield and cuirass against the shield, with long pikes against short ones, with steel swords against those made of iron, with armed chariots against infantry, etc. He excels in inventing rifles and cannons, distancing the battle as far as possible, so he can distance himself from the struggle.

Any human action is the carrying out of an aspiration (desire)—whether intended or unintended. When there is only one aspiration, then it becomes the direct impetus to action, which turns out to be a simple act of will (impulsive). However, the human psyche usually generates several simultaneous desires, which often contradict one another.

The clash among different desires and different emotions, which accompany and induce these desires, constitutes the so-called conflict of aims and motives.

It is impossible to avoid mentioning the significance that emotions have in human activity. While motives are thought through and carefully measured, emotions drop the full weight of their influence on the scale. {Korff, 'On the Cultivation of the Will of Military Commanders'.}[37]

The soldier's aspiration to 'win' drives him even before he enters battle. However, the ambience of battle, with its most crucial element—danger—generates within the human psyche a desire to 'evade danger'. Following our discussion above about how the desire 'to evade danger' changes the ways of conflict, we can summarise this

---

[36] The quote is from Ardant du Picq, *Battle Studies*, p. 43.
[37] Golovin refers to Nikolai Korff, *O vospitanii voli voennochalnikov* [On the cultivation of the will of military commanders], (St Petersburg: Obschestvo Revnitelej Voennykh Znaniy, 1906).

change in the following words: 'to win with the least risk'. However, since danger increases during battle, a significant conflict develops between the two motives ('to win' and 'to evade danger'); and the formula 'to win with the least risk' cannot simply resolve this conflict. **The desire 'to evade danger' becomes almost equivalent to the desire 'to win'**. If we consider the soldier's desire to 'win' a positive motive, then his desire to 'evade danger' should be considered a negative one.

This struggle between two contradictory motives is nothing but a drama. The degree to which this drama is painful for soldiers, we can see in the suicides which sometimes occur during battles, and in cases of intentional self-harm, which are in fact quite frequent.

While participating in battle, the majority of soldiers act mechanically. They spend so much energy on their internal struggle against the instinct of self-preservation that they do not have enough spiritual energy left for any independent activity. These soldiers need to receive an external impulse, a trigger to act. This impulse is usually given by those few brave people who preserve their peace of mind and keep in reserve their spiritual energy.

The soldier's negative desire derives from danger. Since the danger increases as the enemy gets closer, the soldier's negative desire also becomes more and more powerful. When the power of the negative desire overwhelms the soldier's positive desire to 'win', then we can see a crisis of will: the soldier decides to evade battle.

The number of these 'evaders' can reach significant levels, as was demonstrated by Du Picq in one of the examples he analysed: 'Let us take Wagram, where his (Napoleon's) mass was not repulsed. Out of twenty-two thousand men, three thousand to fifteen hundred reached the position ... Were the nineteen thousand missing men disabled? No. Seven out of twenty-two, a third, an enormous proportion, may have been hit. What became of the twelve thousand unaccounted for? They had lain down on the road, had played dummy in order not to go on to the end.' {Ardant du Picq, 'Battle Studies'.}[38]

And these were the soldiers of Napoleon's army, whose fame reverberated around the world. [...]

---

[38] The quote is from Ardant du Picq, *Battle Studies*, p. 150.

While the evaders are weaker people, as long as their brave counterparts hold their ground, the military unit is still alive. These lionhearts can be few, but thanks to them the whole, against which the enemy is fighting, survives. As the enemy cannot see what is happening inside this whole, the battle continues.

However, there may come a moment when this whole, i.e., the whole military unit, refuses to continue the battle; when it lies down and is uncapable of standing up; when it retreats or even runs away in panic; when even those lionhearts, who till then had successfully suppressed the instinct of self-preservation, fail, defeated by this instinct—then the end of the battle approaches.

Here we enter the field of collective psychology, on which the phenomena of battle depend—something that we will discuss later. For now, we will try to answer the question, the wrong answer to which has been the main reason behind the extreme paucity of psychological research among the treatises written by the professionals of military affairs. I am talking about the widely accepted opinion that truth about war can impair the battle-readiness of an army. Is this really so?

3. An Understanding of the 'Internal Side' of War Is Impossible without Objective Analysis of Its Psychological Nature

[…] Some historical examples may suggest that the 'truth' about war can actually weaken an army's spirit. Indeed, how often do doctors hide from their patients the inevitability of the approaching end of their fatal diseases, trying to ease the final days of their lives? I have no doubt that the centuries-old military tradition of embellishing war, presenting it not in its real form, has practical foundations. Moreover, I would argue that in wars between contemporary developed nations, such a primitive device as distorting or hiding the 'truth' about war would be completely useless and, in certain circumstances, could even be dangerous. Indeed, in the 'good' old times of professional armies, when battles were very rare episodes of military campaigning and lasted no longer than a few hours, it was possible to keep the soldiers blind to the true reality of war. Nowadays, however, when wars are waged by whole nations, when they can last for years, and when the battlespace has spread over time and space—is it possible to hide the truth, which will inevitably be seen first-hand by every soldier from the first moment of action? Is there not a danger that loss of spirit, because of shattered illusions, will only increase the moral depression experienced by the mass of ordinary soldiers entering the danger zone? Is there not a danger that under these circumstances soldiers will develop the opposite extreme tendency: to see everything in black? [...]

It would be interesting to recall here the measures implemented in 1917 by the chief of staff of the French Army, General Pétain, to save that army from collapse. He demanded that in this critical moment for the army's spirit, officers remain with their soldiers at all times, maintaining the closest possible contact with them. General Pétain's calculation was right, as in such a situation it is easier for a soldier to get answers from his officer to the questions that torture his soul. [...]

The collapse of the Russian Army in 1917 presents an example to the contrary. No other European army involved in war had so deep and wide a gulf between the perceptions of war predominant in military headquarters and its reality as experienced by the soldiers. With the beginning of the Revolution, owing to general social causes, the gap between the soldiers and their officers started to grow. The gap continued to grow, reaching the level of open hostility after General Kornilov carried out his ill-conceived and unsuccessful action.[39] Left to themselves, Russian soldiers, who were much less educated than their French counterparts, became easy prey for the Bolshevik propaganda. [...]

Therefore, the previously held opinion about the danger of disclosing the age-old truths about war should now be revised, even in relation to purely practical affairs, such as the establishment of armed forces or the training of soldiers.

For example, let's take the field of military leadership. On the modern dispersed battlefield, the direction of forces is almost exclusively based on reports from the field. In the last war, especially at its start, the following situation could be observed. Immediately after the first action, a gap developed between the understanding of the situation by

---

[39] Golovin refers to the Kornilov Affair, an attempted military coup d'état by the commander-in-chief of the Russian Army, General Lavr Kornilov, in September 1917 against the Russian Provisional Government headed by Alexander Kerensky.

the forces on the ground and that by the senior headquarters. While the former reported the truth, the latter thought that the forces had failed to do what they must do and were actually capable of doing. The headquarters established a system of command based on 'demands', a typical example of which is offered by the commander-in-chief of the Northwestern Front, General Zhilinsky,[40] during our first operation in Eastern Prussia. The same occurred in our other senior headquarters, though to a lesser degree. I used to hear too often the following argument: forces always do less than they are ordered, therefore it is necessary to 'demand' more from them, as only then will they do what is required. This fallacy had become so prevalent that many of our senior commanders considered giving gruelling orders as a sign of their energetic commandership. The forces, however, paid an enormous price for this system of 'demands'. Our failures in Eastern Prussia can serve as a good illustration of that. However, the forces on the ground quickly learnt the brutal lessons of reality; they adapted and developed a system of 'conditional', if not 'false', reports. For example: an infantry unit is sent in daylight to attack where there is barbed wire; obviously, the attack is unsuccessful; to avoid another order, calling for a repeat of this hopeless attack, the commander of the regiment reports that the soldiers lay down and cut the barbed wire. He counts on the inability of the senior headquarters to understand the realities of contemporary warfare, assuming that this will limit their capacity to grasp the fictitious nature of his report. Another example: realising that brave and successful attacks were not appreciated by their superiors if the reports were not accompanied by descriptions of mythical clashes of bayonets and cavalry strikes, adroit commanders started to produce reports which followed a pattern that resembled battle pieces from times long past. [...]

These are the consequences of the outdated, antiquated point of view which assumes that hiding the age-old truths about the nature of war will allow one to preserve the highest levels of an army's spirit. Half a

---

[40] Yakov Zhilinsky was a Russian Imperial general. He served as chief of staff from 2 February 1911 to 4 March 1914. At the beginning of the First World War, he assumed command of the Northwestern Front. After the fiasco of the East Prussian Campaign, he was relieved of his command.

century ago, Dragomirov already wrote: 'We believe that it is finally the right time for a sober approach to military events: the Agamemnons, Achilleses, and other more or less beautiful epic heroes must step down from the stage, giving way to common people with their great virtues as well as their sometimes humiliating weaknesses; with their selflessness, which forces them to sacrifice themselves for their friends, as well as their selfishness and self-interest, which forces them to eliminate (at the hands of the adversary) those close friends whom they suddenly dislike; with their ability to climb vertical walls without any help and under volleys of bullets and grenades, as well as their habit of retreating sometimes, and running away without looking back, only because some scoundrel decided to cry: "we are surrounded".

'Military affairs can only win from that: truth always helps one to win. Theoreticians should stop limiting themselves by using the all-excusing phrase "this was an accident", and focus instead on the question of organising forces so they will be less affected by unfavourable accidents.'[41]

An understanding of the 'internal' side of war is possible only after the full truth about war is disclosed. [...]

4. Subjective Psychoanalysis and the Methods for Its Application

The last war offered material unprecedented in size and scale to enable military researchers to carry out a psychoanalysis of soldiers. Those recruited to the forces of the developed nations were highly educated and qualified. Before becoming soldiers, many of them were professors at universities. Well familiar with the methods of scientific analysis, they applied their skills to discuss the phenomena and events of war, in which they themselves directly participated and which they personally experienced.

However, we should not close our eyes to the realisation that the proper study of all these many memoirs and records requires a colossal amount of preparatory work. [...] Only after a meticulous examination of the credibility of each of the memoirs, and the degree of truthfulness of each of the authors, will it be possible to use the chosen material.

---

[41] Dragomirov, *Razbor romana 'Voyna i Mir'*, p. 4.

The subsequent work would need to be divided into several stages, starting with the classification of files. Only the last stage would be able to offer general conclusions.

Yet, these conclusions would require a more objective verification than the simple comparison of the testimony of one witness with that of another.

The first attempt to create this objective method of verification was made by a Russian doctor, Gerasim Shumkov. In the aftermath of the 1904–5 Russo-Japanese War, he collected the testimonies of war participants, conducting a series of research projects into the physiological changes occurring in the human body under the influence of the dangers of battle. Dr Shumkov was correct, seeing in these physiological changes the required objective parameters which could direct an experiment of verification in the field of subjective psychoanalysis.

Unfortunately, Dr Shumkov's attempt failed. Apart from a few experts, no military scientists paid attention to his research. Moreover, in the field of civilian science, his project ended up in a small scandal. When Dr Shumkov was preparing to report his findings at a meeting of our medical experts, they declared that a report about such a barbarian act as war contradicted the very foundations of their civilised organisation. Shumkov's report was interrupted and his work was forgotten.[42]

While the case of Dr Shumkov is outrageous, it is, unfortunately, quite representative. The ugly treatment that it received demonstrates an unwillingness to study war that is so prevalent in the majority of civilian scientific institutions still today. The consequences of this detrimental situation are obvious. For four full years, almost the whole civilised world writhed in agony; however, in the following twenty years, no proper scientific research in the field of military psychology has been conducted.

The classification of psychoanalytic data mentioned above requires such a colossal effort and means that it can be achieved only when at least one scientific institution in the world acknowledges the need for a sociology of war, as a science equal among other social sciences. Only

---

[42] This is not entirely true, as Shumkov successfully published his findings a few years later. See Gerasim Shumkov, *Psikhika boytsov vo vremya srazheniy* [The psyche of fighters during battles], (St Petersburg: B.I., 1909).

after the establishment of this science will it be possible to start any coherent collection, classification and analysis of the material present in memoirs about the last war, as well as the rich experience accumulated during the whole of military history. A metaphorical comparison can best explain this requirement. Contemporary life has become so complex that every large building is erected according to detailed specifications. A block of small flats is built differently from a block of luxury apartments; a perfume factory is planned differently from a car plant. No one would contradict this. Therefore, it should be obvious that the establishment of military psychology requires, first and foremost, a developed plan, which will include all the necessary stages. This plan should make clear that its last stage, which can lead to practical outcomes and lessons, is only possible after the conclusion of all the other stages, which have no direct utilitarian meaning. This plan will be coherently implemented only if it is directed mainly by the sociology of war.

5. 'Crowd Psyche' in the Phenomena of War

So far, in discussing the psychological side of the phenomena of war, we have focused only on the field of the psychology of the individual. Meanwhile, humans do not wage wars as individuals, and thus the data provided by the psychology of individuals cannot offer a full explanation of either the behaviour of the masses or the behaviour of the individuals that constitute these masses. Hence, we must enter the field of collective psychology, despite its extremely poor scientific development. The most significant steps in this field have been made to analyse the laws that govern the so-called 'psychological' crowd.

In its common meaning, the word crowd implies a gathering of individuals, regardless of their nationality, profession or gender, and the reason for this gathering. However, from the psychological point of view, this word has an entirely different meaning. Under certain conditions, a gathering of people has very specific features and characteristics. Therefore, from the psychological point of view, the word crowd means a gathering of people that complies with particular laws.

Without any doubt, the mere presence of many people in one place is not enough to call such a gathering a crowd, in its psychological meaning, as it requires certain conditions to exist.

For example, sometimes six people are enough to create a gathering which, from the psychological point of view, can be called a crowd. However, sometimes even hundreds of people, accidentally gathered together, will fail to constitute a crowd, as they lack the required conditions.

A 'psychological crowd' has its own special features. While these features may be temporary, they are very distinctive. We should not see a crowd as a gathering of a certain number of people whose vices and virtues can simply be added together, even though this mistaken opinion was promoted by Herbert Spencer, who saw a crowd as a sum or, to put it correctly, as an average of the intellectual and moral characteristics of the individuals that constitute that crowd. {Spencer, 'Psychology', Chapter VII.}[43]

In this case, a comparison from the world of chemistry fits better than one from the world of mechanics. Like chemistry, where a combination of two gases can produce a liquid, the characteristics of individuals within a crowd cannot simply be added together. Existing research in collective psychology shows that individuals within a crowd acquire a kind of collective soul, which forces them to feel, think and act as one unified being on its own. Even though these individuals can have very different intellects and characteristics, their actions are conducted in complete harmony. The moral and intellectual characteristics of the individual disappear, his conscious personality is lost, he becomes part of a whole: a crowd that has a life of its own. [...]

In a psychologically unified crowd, the leading unifying element is not the mind, but feelings and instincts. As a consequence, there are three conditions required to accelerate the process of turning a gathering of people (including also an organised gathering, such as a military unit) into a 'psychological' crowd:

1. Destruction of major obstacles that obstruct the unification of individuals into a crowd, i.e., the destruction of their individualities (intellect and will).
2. Unification of individuals by one feeling or instinct.

---

[43] Golovin refers to the Russian translation of Herbert Spencer's *Principles of Psychology*, vol 1, published in 1876.

3. Increasing the susceptibility of individuals to impressions, i.e., the amplification of the interaction of feelings.

The circumstances that surround soldiers in battle recreate these three conditions exactly.

1. Conditions of battle serve to paralyse a soldier's intellectual capacity. This facilitates the disappearance of the conscious personality, which is one of the conditions favourable for the creation of a 'psychological' crowd.
2. The instinct of self-preservation seeks to control a soldier's consciousness. This happens when soldiers' feelings and thoughts are oriented in one specific direction (self-preservation), which is also one of the required conditions for the unification of a crowd into one psychic whole. [...]
3. The internal condition of a soldier in battle is a struggle between two opposite desires. Even where the positive desire finally overwhelms the negative, it also contributes to the complete unification of the crowd. Since only an extremely powerful feeling can overwhelm the very strong instinct of self-preservation, the very victory of this positive desire will lead to the one-directional orientation of soldiers' feelings and thoughts, i.e., their unification into a psychological crowd.
4. The intensity of the internal struggle which soldiers experience in battle tends towards an overwhelming crisis of will: the majority of soldiers are in a state of indecisiveness, their attention is unfocused, their mind is clouded, i.e., they are in a state of complete spiritual passivity. This state reminds one of the psychological state of a person under hypnosis, creating the most favourable conditions for impression. [...]

Regardless of how well they are organised, groups of soldiers strive to turn into a psychological crowd. Moreover, as the battle proceeds and the dangers increase, this desire increases as well. From the moment a group of soldiers turns into a psychological crowd, the period of the so-called 'psychological crisis' begins. [...]

In battle, the ability of a soldier to think for himself decreases under the impact of danger. However, even if it had not been entirely destroyed, in the moment of psychological crisis it falls to zero.

Simultaneously, owing to the increased susceptibility of soldiers to impressions, the power of the feelings receives unprecedented strength. Hence, in these moments of crisis, the battle has a completely elemental character. [...]

8. The Need to Create the 'Psychology of War' as an Auxiliary Science

Everything said above should lead to the following conclusion. The establishment of the sociology of war requires not only significant work in the field of individual psychology, but also, and even more significantly, work in the field of collective psychology. Moreover, the research framework of the latter should include not only the psyche of the 'crowd', but also the whole set of phenomena called 'social psychology'.

What should be the main task in this hitherto unresearched field? The extensive material required for the establishment of this science rests hidden in military history treatises. However, finding it would require meticulous research and classification. [...]

There will, however, be a difference between the analytical methods used by individual military psychology and collective military psychology. While the former will pay more attention to psychoanalysis, i.e., focusing on the 'qualitative' side of the observed phenomena, collective psychology (especially where it focuses on 'social psychology') will pay more attention to the 'quantitative analysis of the phenomena under study'. Focusing on the masses, rather than individuals, it will be able to adopt statistical methods, finding (more often than individual psychology can) 'the laws of large numbers'.

Simultaneously with the development of the military-historical method, the creation of 'special military psychology' will require a series of monographs based on the analysis of the experience of 1914–18. The work on these monographs should begin as soon as possible, while memories of the war are still fresh. A humble attempt of this type was made by the author in his work 'The Russian Army in the World War', published by the Carnegie Endowment for International Peace. {N. Golovin, *The Russian Army in the World War: A Sociological Study*, New Haven: Yale University Press, 1931.}

In the last two chapters of this book, the author attempts to sketch the process of disintegration in the Russian Army that led to

Bolshevism. {The author also presented a paper on this topic ('La désintégration de l'armée russe en 1917') at the 13th International Congress on Sociology in Paris in September 1937. The following can serve as a good example of the extent to which the vast majority of the representatives of general sociology are not interested in the idea of the sociological analysis of war: in the pack of written papers, distributed among the participants, the author's paper was significantly misrepresented and distorted.} The author, however, recognises that it was a mere outline that needs to be significantly expanded by analysis of primary sources, which, in fact, exist in abundance. [...]

War creates conditions in which the activities of peoples of each of the warring nations are highly interconnected. Therefore, any conclusions made in one field of individual military psychology are ultimately doomed to be one-sided.

Moreover, the danger of one-sidedness, which threatens individual military psychology if it attempts to generalise its conclusions beyond its boundaries, also threatens military psychology as a whole. Military psychology analyses the phenomena of the psychological side of war, which is rooted not only in the spiritual dimension, but also in the material one. The interconnectivity between the spiritual and material sides of each phenomenon of war is so close that they are organically **inseparable**. For example, better military equipment increases the morale of the army and simultaneously reduces the morale of the opponent; awareness of significant numerical advantage has a similar effect on the morale of both sides.

Unfortunately, too often we see military writers who, blinded by their desire to highlight the importance of the spiritual element in war, miss this close and inseparable interdependence between the spiritual and material sides of the phenomena of war. Even as distinguished a military scholar as Dragomirov did not escape this mistake. To prove the superiority of the spiritual element in the army, he contrasts the spiritual element with the material one: a brave soldier with a poorer weapon versus a cowardly soldier with the best weapon. The mistake of this comparison is that the best possible weaponry does not necessarily correspond to cowardice. On the contrary, as we said above, the existence of better weaponry (and the skills required to operate it) not only increases the morale of the forces, but also reduces the morale of the enemy. The result of this mistaken assumption was the following paradox: we who, before 1914, had talked more extensively about the superiority of the spiritual element than the Germans, entered the war with artillery support for our divisions that was twice as weak as that of the Germans—hence, we lowered the morale of our troops, and reduced the belief in our invincibility.

This is why the conclusions of military psychology alone cannot be considered final and definitive, even though it analyses the most important aspect of the phenomena of war. Such conclusions can be made only when the spiritual side of the phenomena of war is reconnected to its material side. If we take the conclusions, achieved by military psychology, as a thesis, and the conclusions, achieved by analysis of the material side, as an antithesis, then the final conclusion can be achieved only through synthesis (generalisation) and **not through contradiction**.

As a result, military psychology, even in its broadest terms, can only be an auxiliary science at the service of the sociology of war, which is the highest synthesising science of war. Therefore, if the creation of the sociology of war requires the fastest possible creation of the psychology of war, then the latter can get proper direction only when the sociology of war is sufficiently developed. [...]

CHAPTER FIVE

CONCLUSION

While the first chapter sketched the main aim of the sociology of war, the following ones focused on the questions regarding the material and methods required for the creation of this new science.

From these chapters, it becomes obvious that the main difficulty of this creation is the colossal work required to collect and analyse the vast material assembled by humanity for entirely different purposes. This process is so complex that its successful accomplishment requires not only an analysis of the materials collected by military history, but also the introduction of new research methods to military-historical science itself. Moreover, it also requires the simultaneous introduction of auxiliary sciences, such as the psychology of war and the statistics of war.

These tasks, however, require the existence of the sociology of war as a precondition for their achievement, thereby creating a chicken-and-egg situation. The only possible way out of this situation is simultaneous work in all the fields mentioned above. The collection, classification and systematisation of material must occur simultaneously with the development of the methods, as well as the sociological analysis of war. [...] Finding the most productive way to collect material, choosing the most correct classification and registration of this material, developing the most suitable methods to analyse this material—all these require at least some knowledge of the features and characteristics of the subject being analysed, which can be achieved only through preliminary research. It doesn't matter if the fate of this research is similar to the fate of scaffolding. Although the scaffolding is removed when the construction is completed, it is absolutely necessary for the building to be erected. This method, which can best be called the method of 'working hypotheses', has been used successfully in all complex sciences. Therefore, there is no reason to look for other methods for the sociological analysis of war. However, it is important to state that the extreme complexity of the internal structure of war does not allow for the creation of a working hypothesis that would embrace all phenomena of war. It needs to take a significantly slower way—a way of building partial working hypotheses, leading to the application of the monographic method, i.e., a thorough analysis of each of the main phenomena of war. [...]

Coming back to the metaphor of scaffolding, it is important to remember that scaffolding can be useful only if the construction of the building occurs simultaneously with the scaffolding's installation. Therefore, the main value of 'monographic analysis' as I suggest will be not as stand-alone research, but in connection with the comprehensive work involved in the creation of the sociology of war and other related special sciences.

Such work can be conducted only by a collective effort guided by a meticulously developed plan. [...] Therefore, in the conclusion of this treatise, I can only repeat what I said at the end of the first chapter.

Only the establishment of a special scientific institute or, at a minimum, a department of sociology of war, at least at one university that has a faculty of social sciences, can be considered as the first true attempt to study war scientifically as a phenomenon of human social life.

NIKOLAI GOLOVIN

While it is difficult to expect that this will bring humanity closer to peace, as pacifists so strongly advocate, one thing can be said for sure—walking on the edge of the abyss with open eyes is significantly less risky than walking blindfold.

6

THE FACE OF CONTEMPORARY WAR[1]

Evgeny Eduardovich Messner

Introduction: Styles of War

The Book of Genesis tells us: '... God saw all that he had made—and it was very good! There was evening, and there was morning, the first day.'[2] And God arranged life on earth with an incomprehensible wisdom, by basing the development of life on the struggle for existence. However, while giving his creatures rule over life, which includes killing, God prohibited unjustified extermination. There are only two creatures that violate this prohibition: weasels and humans. A weasel

---

[1] This chapter is based on the translation of selected chapters from Evgeny Messner, *Lik sovremennoy voyny* [The face of contemporary war], (Buenos Aires: South American Division of the Institute for the Study of the Problems of War and Peace named after Prof. General N.N. Golovin, 1959).

[2] This passage, in fact, speaks about the sixth day, not the first: 'God saw all that he had made—and it was very good! There was evening, and there was morning, the sixth day' (Genesis 1:31). It seems that Messner intentionally changed the day in an attempt to strengthen the point of his statement, which is based on a paraphrased version of this passage at the end of the introduction.

kills more than it can eat and it becomes frenzied in its purposeless killing. A human also at times kills more than the struggle for existence requires, kills because of his hatred, which comes not from God, who is disgusted by it, but from Satan, who enjoys it.

War is one of the forms of existential struggle, and as long as other forms do not eliminate it, it will be allowed by the Law of Life. Yet, not all forms of warfare are allowed, especially those that are reminiscent of a weasel's behaviour.

Each phase in the development of human morality has been accompanied by a different style of war. During the time of knighthood and chivalry the battlefields were like tournaments, where the extent of bloodletting was carefully regulated. Frenzied fratricide, demonstrated best by the St Bartholomew's Day massacre, characterised war in times of religious zealotry. The French Revolution opened a new chapter—the despotic love of freedom—epitomised by Austerlitz and Borodino, where the bloodshed surpassed the terrible slaughters of the hordes in antiquity.

The 19th century was marked by a humanitarian liberalism that tried to tame war: Tsar Alexander I preached humanitarianism; Tsar Nicholas II convoked the international Hague Peace Conference; the Red Cross was created. The 20th century, however, introduced the age of militant pacifism: by waging wars, pacifists try to eliminate militarism and in the name of humanism they conduct their wars in an extremely inhumane style.

The style of contemporary war is very similar to the style of the weasel—extermination. Like the Inquisition which tortured people *ad maiorem Dei gloriam*, pacifism extirpates people *ad maiorem hominis gloriam*. From thousands of years of experience, pacifism has learnt only one lesson—'in war everything is allowed'—completely disregarding the honourable military traditions. [...]

In the 1754 Battle of Fontenoy, Lord Charles Hay called to the French commander: 'Order your men to fire!' 'After you, English gentlemen!' replied the French commander, and the subsequent first volley of the British killed the first line of French troops.[3] This is how

---

[3] Messner refers to the famous story that occurred during the 1745 Battle of Fontenoy.

we used to fight in old times. Nowadays, however, pacifist Churchill condemns the whole German population to death from hunger. He also terrorises German civilians by bombings, while pacifist Einstein delivers into the hands of pacifist Roosevelt a bomb which 'pacifically' massacres the population of Hiroshima and Nagasaki.

Amateur strategists, who are ignorant of military affairs but very skilful in inter-party political struggles that shape their flexible sense of conscience, clutch at Ludendorff's 'total war', turning wars into a matter of total brutality. The term 'total war' was initially meant to describe a war waged by all efforts. It meant that not only soldiers and sailors should fight for victory, but also the locomotive drivers and factory workers, miners and civil servants—in other words, all are obliged to contribute to the war effort; and all should be sacrificed for victory—the wealth of the rich and the destitution of the poor, the knowledge of the scientists, the talent of the writers, and the Samaritanism of women.

The term 'war by all efforts', however, was quickly replaced by 'war by all means'. [...] For thousands of years, wars were wars: soldiers slashed, stabbed or shot one another while carrying out their harsh duty. Russian soldiers fought against Turkish soldiers for two hundred years, but neither the former nor the latter hated their enemy. Nowadays, however, it is not honour that leads our soldiers into battle, but hatred; and our strategists are inspired by satanic hatred that leads them to wicked, non-soldierly actions, which are called wars by force of habit.

The world is full of hatred in times of war, as well as in times of peace. The driving force of the politics of peace, the politics of coexistence and the politics of conflict is hate: racial hate—nations of colour hate white people; hate of inequality—the poor hate the rich, the rich hate the super-rich; doctrinal hate—between the red, the pink, the pale and the white; and, finally, hate for the sole purpose of hate—the hate of evil for everything good or even everything that just seems to be good. [...]

This all-embracing hatred creates paradoxical contradictions between the materialistic consciousness of people and societies and their ideological inclinations. In these times of materialism, societies still take up arms and fight for the triumph of certain ideas. But today, all types of materialism—dialectical, existential, capitalistic, etc.—all have become different types of ideologies that seek to offer omnipotence, not only in the material world, but also in the spiritual one.

STRATEGIYA

In our times of internationalism, which leads towards cosmopolitism, nationalism turns into an irresistible chauvinism, in which at the same time as constructive movements, such as Pan-Europeanism, Pan-Americanism and Pan-Arabism, there are powerful destructive movements expressed in political and armed struggles between societies for their own national pretensions.

In our times of godlessness, religious wars have become possible, because godlessness itself has become a kind of religion. While those following the word of God are ready to fight and die for their 'I believe', others are ready to die with a similar fury for their 'I don't believe'.

The French socialist Jean Jaurès used to say: 'war is a barbaric form of progress'. Both World Wars assisted the progress of those who did not fight—the nations of Asia and Africa were freed from their colonial dependency. For the participating nations, however, these wars caused material and moral regression, regardless of whether a nation was defeated or was a victor itself.

The Scythian tactic of scorched earth, revived by the Unionists during the American Civil War, has been executed with a scientific-technological persistence since 1918 when, retreating to the Siegfried Line, the Germans destroyed everything that could be destroyed by hammer, axe or dynamite. In the Second World War, retreating from the continent, the English destroyed ports in the Channel; and in Burma, unable to face the Japanese offensive, they burned half of the country. In 1941–42, Stalin ordered the ruin of territories before they were occupied; two years later, Hitler ordered the same for the same territories when he retreated. In 1945, five million avengers from the USSR stormed Europe, five million people on the rampage, five million pairs of predatory hands, five million throats thirsty for schnapps, five million depraved souls lusting for women.

When the next war comes, a motorised Genghis Khan will emerge from Eurasia, his intentions having already been glorified in Bolshevik songs: 'We will burn Rafael, destroy the museums and trample the flowers of art.'[4] The West will face this destruction with its own rich history of the barbaric obliteration of Monte Cassino, Dresden and

---

[4] This line is from the poem 'We' written in 1918 by a Bolshevik poet, Vladimir Kirillov.

Hiroshima. This war will destroy hundreds of millions of lives, life itself, and many cultures created by dozens of generations.

The ideological delirium of our times and the all-embracing hatred have shaped the style of contemporary war as something wicked, extremely bitter and apocalyptic. And therefore, it will be written somewhere: 'God saw all that they had made—and it was very bad! There was evening, and there was morning, the last day.'

I. Dimensions of War

From the beginning of time, war occurred in the two-dimensional space of land and, sometimes, on the similarly two-dimensional surface of the sea. The theatre of war was measured by its length and width. In this century, however, war has acquired a third measurement, on land—altitude—and in the sea—altitude and depth. Owing to their secrecy in times of war, underwater operations endanger not only naval seamen, but also non-military ones as well as seafarers, who have no part in war at all: the scourge of war extends to those who try to avoid it.

Air operations have further increased the destructiveness of war. In the two-dimensional theatre of war, a strike can be parried either by a counter-attack or by fortification. In the aerial dimension, however, the meteoric speed of the strike significantly restricts counter-attacking options, and averting it by means of fortification is simply impossible. [...]

At all times, the populace felt secure behind its military, if it trusted it. Today, there is no reason to trust land or air forces to secure the population: they are powerless to prevent enemy air force strikes on the home front. The word 'front' has lost its previous meaning, as it now threatens the state as a whole in times of war. The word 'soldier' has also lost its meaning of courage, as courage is also required of women, who have to perform their daily peacetime duties under the terrifying sirens that forewarn of approaching bombings. [...] The deprivations of war, suffering and danger have become the fate of all people. [...]

As long as the fashion for 'total war' did not consume the last bits of strategists' conscience, they restricted themselves to bombing military targets only. But while their conscience was shrinking, the understanding of 'military target' was extending, covering everything—the totality of war has become the totality of targets—as even a button factory and a milk farm have been defined as 'war industries'.

'Total war' has turned into total destruction, total killing and total madness. Pompeii was destroyed by the power of nature, and its destruction has been remembered for centuries. Today, the power of human beings destroys hundreds of Pompeiis, but these 'events' are so trivial that humanity barely pays attention and will definitely forget them in ten years' time.

Our wars of 1854, 1877, 1904 and (partly) 1914 were duels between armies—the population was in the position of seconds. Today, however, the duel has been replaced by total slaughter: with the help of aviation, war has been transformed from a matter of surface into something that has volume: it acquired a third dimension, it has become a massacre.

While the third dimension turned war into a massacre, the most tragic transformation occurred after the development of the fourth dimension. The weakening of the enemy's military and people's spirit has always been a subsidiary effort of war. [...] Nowadays, however, there is a methodological approach to this issue: the soul of the enemy's society has become the most important strategic objective. [...] Degrading the spirit of the enemy and saving your own spirit from degradation—this is the meaning of struggle in the fourth dimension, which has become more important than the other three dimensions.

While the splitting of the atom for the purpose of massacre was carried out for the first time in 1945 (Hiroshima), German scientists had started to work on the splitting of an atom for scientific purposes forty years previously. Similarly, while the first large experiment on the splitting of the spirit occurred in 1917 (Woodrow Wilson's fraudulent declaration with its fourteen points),[5] the first systematic approach to the splitting of the national spirit occurred many years before, in 1864, when Marxists established the International Workingmen's

---

[5] Messner refers here to the famous 'Fourteen Points' speech on the war aims and peace terms of the First World War delivered by US President Woodrow Wilson on 8 January 1918.

Association in London. The slogan 'Workers of the world, unite!' revolutionised human life more than nuclear physics revolutionised life by transformations in technology, industry and war.

The idea of the national state was created at the end of the 18th century by the French Revolution, but already at the beginning of the 20th century the vertical barriers between national states started to rot. Instead of these rotting barriers, new ones started to appear, those that divide humanity horizontally between bourgeois and proletariat; Marxists, democrats and Fascists; Catholics, Muslims and atheists.

In the last war, the fields of the Soviet Union were overrun by those who marched with the Führer's Germans contrary to their national spirit: the Spanish 'Blue Division', as well as divisions from Italy, Holland, Hungary and Romania; the 'Wiking' division recruited from Danish, Norwegian, Swedish, Finnish, Dutch and Belgian volunteers; and 'Free Corps Denmark'. This is how Fascist internationalism overcame the vertical national barriers in its war against Marxist internationalism. [...]

A tiny but very representative example of the splitting of the national spirit occurred in Serbia: rejecting the covenant of their ancestors, 'Only unity saves the Serbs', people turned to a bloody conflict: Nazi general Nedić, democratic colonel Mihailović and Communist bandit Tito tore Serbia's body and soul apart. [...]

In the past, there were revolutionary wars and military revolutions, but today these two calamities are interwoven in war-revolution, i.e., four-dimensional war. Today, a less-than-four-dimensional war is simply impossible. It is possible to imagine a large-scale war that is not nuclear, but it is impossible to envisage a war that does not employ the splitting of the spirit.

Nowadays it is easier to degrade a state than conquer it by arms. National states have become morally vulnerable, because the spiritual meaning of the state has weakened. [...] The people's repulsion from the power of their own state is the feeling that fires the struggle (from outside and from inside) against the hated regime—this is the 'uranium' of the spiritual splitting 'industry' which has become enormously important nowadays.

The infiltration of Lenin into Russia boomeranged on the Germans. Any operation in the fourth dimension can boomerang, though no one can withstand the temptation of using underground forces inside the enemy's camp, even if these are Beelzebub's forces.

The fourth dimension is an integral part in these times of confused minds and absent conscience. And this is very tragic because, if the third dimension is like the infliction of all Ten Plagues of Egypt together, there is only one word that defines the fourth dimension of war: Hell.

II. The Chaos of War and Peace

The twenty-storey building of the United Nations in New York, without any balcony or ledge, observes the world dumbly with its five thousand identical windows. In return, the world sees in this box a symbol of the equality of states, as all states in the United Nations are equal: tiny Lebanon and huge Indonesia, newly created Libya and ancient Italy. Even the privileged 'Great Five' can express their will only by a passive 'don't allow' rather than an active 'I wish', and to carry out its plans, all-powerful Washington must recruit the votes of tiny states that have learnt to appreciate their value in this new global situation.

The time of dictatorial states in global affairs has passed. [...] The time of isolated hegemons has passed too. Today, hegemony can only be the right of coalitions created on the basis of mutual exploitation and reconciliation of interests. [...]

The hegemony of Europe was undone by two unnecessary events of European conflict, which turned into world wars. [...] Europe, which considered itself the centre of the world from antiquity through the Renaissance, the Reformation, the Enlightenment and the French Revolution to the humanism of the 18th century and the liberalism of the 19th century, today has become the back of beyond. [...]

Exhausted by wars, Europe is afraid of any decision, and it is resolute only on one issue—its refusal to rule the world. [...] Europe has no centre, nor is it the centre of the world anymore. The centre of gravity of civilisation is shifting from Europe. [...] It is difficult to say where the centre is shifting to, but nowadays there are only two main poles: the United States and the Soviet Union.

Anyone who speaks about war these days implies a conflict between these two great superpowers. Because, nowadays, a real war is possible only between great powers, and an armed conflict between less-than-great powers is as meaningless as war between Bolivia and Paraguay was a quarter of a century ago. Local conflicts today are just 'incidents', but a 'war' is a global or near-global phenomenon, because there is little chance that someone will remain neutral in the contemporary interwoven knot of interests.

In the 15th century, Portugal and Spain significantly increased their power by the colonial conquest of new territories, but it created little jealousy in France or in the Holy Roman Empire. Nowadays, if Austria tried to forcibly take away from Italy an Alpine valley inhabited by Austrians, the whole of Europe and the whole of the world would intervene in the conflict. Some would do it to preserve the balance of power, others to seek a balance of principles; some would intervene owing to economic reasons, others owing to personal ambitions—why should I not trumpet my opinion when everybody else does?

War in our days can only be waged by a coalition: by a limited coalition or an unlimited coalition, i.e., a world war. Even in a local war, there would be some overt or covert shareholders: behind North Korea there was Red China and (less obviously) the USSR, and behind South Korea there was a huge coalition, though the majority of the partners offered only symbolic blood sacrifices on the altar of the common goal (the delivery of tinned blood for transfusion does not count), but all participated in strengthening the hands of the strategists.

This type of coalition war has neither clearly defined goals nor clearly crafted strategy, because there are too many interested players, whose interests are not only difficult to negotiate, but are sometimes completely opposed to each other. Churchill's interventions in the Dardanelles campaign derailed the Russian plan to land at Constantinople, because the Russian desire of opening a sea link to its allies contradicted Britain's traditional policy of not allowing Russia to overthrow Turkey. In the Second World War, Churchill again struggled with Roosevelt, wishing to see Eisenhower in the Balkans and not Tolbukhin.[6] Such 'internal fronts' are an integral part of any coalition, even though fragmentation of the primary and secondary goals of war ultimately leads

---

[6] Fedor Tolbukhin was a Soviet general. During the 1944 Summer Campaign, he was in command of the Soviet invasion of the Balkans.

to a vague and indecisive strategy and, therefore, to a slow unfolding of events and the prolongation of war.

In the past, wars were waged for the purpose of conquest or liberation, for the purpose of looting or taking revenge on those who had looted us before. Nowadays, wars are fought for 'a place in the sun'. [...] The world has become too cramped for humanity. Asia and Africa press themselves on a Europe already full to the limit. America, exactly four hundred years after the Cortés invasion, invaded Europe in return, using its soldiers, and then the remaining continents, using its finances and diplomacy. [...]

Hundreds of millions of people think that they are under the threat of a Communist invasion, while several other hundreds of millions are afraid of retaliation. Hundreds of millions suffer under the yoke of Communist dictatorship, while many other hundreds of millions imagine that their mission in life is to fight colonialism, which, in fact, has already disappeared from existence. Many countries are in a constant fully mobilised or half-mobilised state of readiness, i.e., half of the American strategic air force is constantly up in the skies, fearing to become the victim of a Soviet strike. The outbreak of classical regular and contemporary irregular wars unnerves humanity, because nobody feels safe from a war-revolution or a revolution-war.

Decades will pass and humanity will cure itself of the current madness. The problem is that there may come a moment when the 'grace' of this world will become unbearable for people and they will dream about war, attempting to use a temporary evil to eliminate the prolonged evil of that tragic curiosity—a world that is neither at war nor at peace.

There was a time when generals thought that all international illnesses could be cured by war, and civilian leaders assumed that it might cure only certain diseases. Nowadays, however, no one thinks so anymore, because the experience of the last two wars teaches that by killing the bacterium, the medicine damages the organism. [...]

It is not surprising that Eisenhower was more peace-minded than Truman, because, as a general, he understood better than a politician the kind of apocalyptic chaos of future wars. Wars have always been chaotic and generals have always tried to curb the chaos by making plans and giving orders. The places where generals failed were determined by military historians; therefore, in books, wars have always looked very neatly groomed, pretending to be in order and not in chaos.

Nowadays, a war would be boundlessly chaotic. There are many reasons for this, such as the exacting variety of different means of war, the multiplicity of goals, the complexity of intra-coalition relations, the difficulty of coordinating operations in the four different dimensions, etc. [...]

The military machine has become too complex and cumbersome. This is why the generals of our times have become amilitarists—they do not oppose war or the military, as anti-militarists do, but rather support the solution of political questions without using military force, at least as long as it is possible. Alas! It is not possible. There will be wars. It is just important to realise that contemporary war exhausts the defeated side no less than the winning one, and that in eliminating several causes of international tension, war creates new causes for new tensions. If in the past the decision to go to war was made in the hope of, or even confidence in, a victory, nowadays it is likely that people will rush into a fight telling themselves: 'It might be worse, but at least it's different!'

III. War and Half-War

Every phase in human development has created a certain form of armed struggle. In medieval times the army consisted of nobles' retinues, which were autonomous in their relations with their leaders just as nobles were autonomous from their sovereign. In the age of absolutism, the army became royal, and the main way of recruitment was selective conscription.[7] When all members of society received their rights (after the French Revolution), the army became national, based on the idea of compulsory military service.

Nowadays, parliamentary democracy progresses towards direct popular sovereignty: riotous crowds dictate their will in the West, and

---

[7] Messner refers to the era of Peter the Great, which symbolises the establishment of absolutism in Russia, and Peter's military reforms in the late 17th century, which first introduced the conscription of peasants and townspeople based on a quota system per settlement.

in the East workers' resolutions create the fiction of a people's government. Corresponding to this departure from the historical division between rulers and ruled, another traditional division has disappeared—that between soldiers and citizens. This leads to the development of a new definition—the citizen-soldier—as every citizen has the right and obligation to participate actively in an overt or covert war.

Today it is impossible to imagine a war without 'resistance', without an underground 'home army',[8] without partisan units. Today we have to reckon with the fact that there is no longer a division between the theatre of war and the country at war; the entirety of both adversaries' territories is now the theatre of war [...] Today there is no division between the army and the population—all participate in war with a different relative intensity and persistence: some fight openly, others secretly, some continuously, others only at a convenient opportunity. Today, regular forces have lost their military monopoly: irregular forces, supplemented by underground movements, fight together with regular units or, at times, even more than the regular forces.

This ultimately leads to three different developments. The first is that war has been transformed into a new form. The Algerians devastated France by setting fire to oil facilities and executing civilians in the city streets[9]—this is the contemporary alternative to conducting a cavalry raid into the adversary's deep rear. In Cyprus, British soldiers orchestrated 'Bartholomew's days', terrorising the EOKA terrorists—this is the contemporary alternative to a counter-attack on an irregular front.

Military leaders did not foresee this development, and failed to understand it even when they saw it. The Germans were powerless against Voroshilov's partisans, Tito's bandits, Sikorski's terrorists and de Gaulle's *maquis*.

The second development is that the prevalence of irregular forces has led to a vulgarisation of the term 'military' and, therefore, to the

---

[8] Reference to the Home Army, the dominant resistance movement in Poland during the Second World War, which was loyal to the Polish government-in-exile.

[9] Messner refers to the actions of the Algerian National Liberation Front (FNL) that took place in Algeria and not in France.

decline in military ethics, which had always promoted stratagems, but never cunning. Nowadays, however, military headquarters develop covert ways, promoting strategy, operations and tactics of infamy: terror, perfidy and treachery. Leonidas' force was destroyed at Thermopylae as a result of betrayal; Prince Skopin-Shuisky was poisoned by a traitor; Grant and Sherman carried out various terror activities during the American Civil War—but all these were just episodes. Nowadays, cunning has become systematic and the Arab way of war—'dirty war'—has become the main part of any war. Feeling their powerlessness in Algeria, Massu's[10] officers feverishly started to adopt their experience and practices from their lost 'improper' war in Vietnam—something that Massu's paratroopers had already comprehended, while other French officers had not. Moreover, it is even less comprehended by officers in other countries or by Western political leaders, who have completely failed to grasp the scale of the phenomenon.

The third development is that war-fighting states have become very tenacious. On the one hand, when, in the Battle of Sedan, France lost its army and failed to organise enough forces for further resistance, Paris was forced to ask for an armistice. On the other hand, during two weeks in September 1939, Poland lost all its armies but it continued fighting until 1945.

Previously, it was almost a rule that a country capitulated after its adversary conquered its capital. In Yugoslavia, however, the Germans conquered five capitals: Belgrade, Zagreb, Ljubljana, Skopje and Cetinje—and yet war continued. Previously, after being ratified by the government, capitulation always meant an end to war. However, while Marshal Pétain ratified France's capitulation, part of the French population, led by General de Gaulle from abroad, continued to resist.

In the past, states either lived in peace or waged wars; there was no third situation. This third situation was created by Trotsky: neither war nor peace. This formula of refusal to make peace has now received a new meaning: refusal to wage an open war. The obvious and defined

---

[10] Messner refers to Jacques Émile Massu, who was a French general in the Second World War, the First Indochina War, the Algerian War and the Suez Crisis. He led French troops in the 1956–7 Battle of Algiers, first supporting and later denouncing their use of torture.

line between peaceful and military international relations has been erased. It is possible to be only nominally at war and even forget about it: Andorra had been 'waging' a war against Germany from 1916, and only in 1958 recalled that it had not signed a peace agreement. It is possible for countries to be at war without openly fighting each other: Greece and Turkey have been waging an indirect war to control Cyprus by supporting EOKA and Turkish Cypriots respectively. It is possible to send troops to conquer without being at war: Syrian soldiers secretly infiltrated Beirut to participate in the Islamic–Christian feud, thus trying to conquer Lebanon for the United Arab Republic. It is possible to wage a war while keeping peaceful relations—this is how Turkey fought against the Soviet Union in Korea. It is even possible to wage a war against a coalition and still be a member of it—this is how the Soviet Union, being a member of the United Nations, fought in Korea against that very same organisation. [...]

Nowadays, there are four different forms of international relations: war, half-war, aggressive diplomacy, and diplomacy.

War is an open struggle with weapons. The type of weapons is unimportant, whether the war is regular or irregular. It is also unimportant whether diplomatic relations are withdrawn or not. This is how Israel, which since 1949 has never established diplomatic relations with Egypt, invaded the Sinai Peninsula in 1956, and France and Britain invaded Port Said, declaring that it was not Egypt they were fighting against.

Half-war is a covert participation in a war or feud. Red China standing behind the Viet Minh fought against France in Vietnam; Washington waged a half-war against Peking by manoeuvring the Seventh Fleet in the Formosa Strait; Egypt fought against France in an indirect way when it fuelled the feud in Algeria as revenge for the French attack on Port Said.

Aggressive diplomacy is something that is commonly and foolishly called 'cold war', and equally frivolously could have been called 'hot diplomacy'. Aggressive diplomacy is an enforced form of diplomacy, just as a half-war is a weak form of war. For example, aggressive diplomacy took place when the Kremlin hampered the pro-Washington course of Argentinian politics by staging political strikes, riotous demonstrations and acts of mass violence. The difference between half-war and aggressive diplomacy is that while the former is waged by armies, partisans and subversive groups, the latter is conducted by political means and methods, though some exchange of bullets and explosion of bombs may occur to increase the overall political effect.

The final and fourth form is diplomacy, or political activities, i.e., when the gloves stay on, based on traditional methods of persuasion and threats, begging and extortion. [...]

These four forms of international relations are bizarrely interwoven in this nervous and alarmed world of ours that fears war and doubts the value of the famous aphorism 'War is evil, but it allows the avoidance of bigger evils'. What is the bigger evil nowadays—the unbearably tragic peace or the ferocious war? 'Rejoice at war,' Goebbels used to say to the Germans, 'peace is more frightening!'

IV. Diplomacy and Strategy

Recently, a high-ranking French diplomat said to a high-ranking American general: 'War is too important to be left to the generals,' to which the American replied: 'And peace is too important to be left in the hands of civilians.' Both are correct. When a vessel heads to a destination defined by its owners, they have to consider the opinion of the captain (the strategist), who navigates in the best possible way through a storm; at the same time, struggling with the storm, the captain should not forget that the vessel should arrive at the port designated by its owners (the government). In 1945, however, the world was rebuilt regardless of the logic of strategy—what will the West's strategists do if their diplomats ask them to defend Berlin's nonsense with weapons? It was the right strategy to land Eisenhower's army first in Africa and then in Italy. But, once Rome was taken, strategists persisted in taking the rest of the Apennine boot from Hitler, despite the logic of diplomacy suggesting another solution: to transfer forces to the Balkans so that King Peter, and not Tito, would control Yugoslavia after the war.

In the past it was simple: Kutuzov was tasked to defeat Turkey and to sign a peace agreement with that country at his discretion, as strategy, rather than diplomacy, had the final word. Clausewitz defined an opposite principle: the aims are defined by diplomacy, and when diplomacy has completed its work, then it is strategy's turn to decide. In this way, the primacy of diplomacy was established. But now the Clausewitzian aphorism 'Strategy is a continuation of politics (i.e., diplomacy) by other means' is outdated. It is obsolete because the clear distinction between the phase in which diplomats do their work and the phase when strategists do theirs has been erased. The line between peace and war has been erased. There is no more interchange: peace—war—and peace again. Peace is interwoven with war, war with peace, and strategy with diplomacy.

What should we understand as 'peace'? What sort of peace is that which is called 'cold war'? What should we understand as 'war' if wars are waged not only by the battles of strategists, but also by the competition of diplomats, who manipulate the 'goals of war' (for example, while some trump with the Atlantic Charter, others declare the New European Order as their trump card). In such a state of international relations, diplomats and strategists should not compete for leadership. The aims of the state should be shaped by national ideals and the state's interests, taking into consideration the capabilities of diplomacy, strategy, politics (internal and foreign) and the economy. Some supreme governmental entity should interweave diplomacy and strategy into one bowstring that will shoot both diplomatic and strategic arrows into the very same target.

It is not difficult to imagine such an arrangement. In peacetime, diplomats try to achieve a state's goals by using traditional means: declarations, conferences, intrigues, blackmail, bribery; in the Middle East it would be black coffee that hides the taste of poison and in France it would be a pistol that kills an allied king. Simultaneously, strategists offer diplomats different 'tools of diplomacy', such as military exercises and manoeuvres close to the adversary's border, such as the movement of the US Sixth Fleet in the Mediterranean Sea. When the path of diplomacy comes close to a cliff, at the bottom of which is war, then the opinion of the strategists becomes the most important one—should we risk jumping off the cliff? And, if yes, what should be the exact place and moment? Diplomats should not push strategists to begin military activities in unfavourable conditions, as an inauspicious initial strategic situation cannot be improved even by the most successful operations. The victorious encirclements of 1941 and the successful operations of 1942 did not reverse the strategic ineptitude of Operation Barbarossa, initiated by Hitler and his party's political-diplomatic mediocrities. [...]

In the modern state of international relations, diplomacy very easily turns into aggressive diplomacy: without interrupting (seemingly peaceful) coexistence with an unaccommodating state, it mobilises oppositional forces and revolutionaries inside this state by means of propaganda and bribery. [...] Sometimes, when the diplomatic office is not enough to transfer weapons to underground movements, diplomats ask help from strategists, who reinforce revolutionaries with instructors and techniques of terror and subversion (from their arsenal of irregular forces, which any well-prepared state has as part of its military). [...]

When aggressive diplomacy turns into a half-war, the role of strategists also changes: from being mere advisers to diplomats, strategists start to lead the action. In a half-war, strategists do not operate with forces but with saboteurs, partisans and (in the worst-case scenario) 'volunteers', like the Red Chinese operatives who were deployed in Korea. During the Spanish Civil War, Germany ('the Condor Legion') and Italy ('the Corps of Volunteer Troops') waged a half-war on Franco's side. On the opposite side, the USSR, France and Britain did the same, supplying José Miaja with 'international brigades' and secret service operatives. On the one hand, during a half-war, diplomats become the main advisers to strategists, ensuring that their actions do not contradict major diplomatic goals. On the other hand, diplomats still remain proactive, coordinating their activity with the strategists and continuing to organise the oppositional public inside the enemy state.

The fourth form of struggle is war, and indisputably this remains in the full custody of the strategists, as only strategy can answer the question whether a war can achieve the state's goals when diplomacy and aggressive diplomacy have failed. Strategy also defines the moment that is most opportune for beginning a war—but it does so not with the help of astrologists and horoscopes, as Hitler, who imagined himself a strategist, did, but with the help of a comprehensive strategic analysis. Such an analysis cannot be done by any general or colonel of the General Staff [...], and definitely not by members of parliament, ministers or dictators, even if they had climbed up the military ladder during their national service. [...] Diplomats should correct the plans of strategists in times of war, when their operational actions disturb diplomatic achievements; but, in a paraphrase of Pushkin's words, 'Shoemaker! Judge things no higher than the boot,'[11] diplomats and presidents should not rule over strategists in the middle of military actions. [...]

While this is an ideal arrangement of the four steps of international struggle, in practice everything looks quite different. In peacetime, parliaments jealously monitor the military, being fearful of the creation of a 'military junta', which, owing to the so-called bloodthirstiness of generals, would force the country to the edge of war. However, those same parliaments and other different types of political actors create non-military juntas, which direct diplomacy without considering strategists at all. Such non-military juntas jump through the hoops of diplomacy until they find themselves in a deadlock, commanding the strategists to begin battle, regardless of how ridiculous this sounds in strategic terms. In 1940 Italy was a threat, considered by France and Britain as a significant danger in the Mediterranean; however, once Mussolini and Ciano rushed to win their laurel wreaths, Italy turned into a joke: there was no suitable terrain to deploy the army against France and there was not enough naval power to fight the Royal Navy.

Moreover, in aggressive diplomacy diplomats and political leaders can proceed too far without being advised by strategists. The Egyptian dictator Nasser escalated the situation in Lebanon to the extent of a half-war: he sent Syrian 'volunteers' to overthrow the Lebanese gov-

---

[11] Aleksandr Pushkin, 'Shoemaker (A Parable)', 1829, translated by Yevgeny Bonver, https://www.poetryloverspage.com/yevgeny/pushkin/shoemaker.html (accessed 20 July 2020):

> Once a shoemaker, on the art's creation,
> In drawn shoes had found a mistake;
> With his fast brush, an artist made correction;
> But the shoemaker went without a break:
> I think the face a little crooked is shown ...
> The breast's much bared, as I've understood ...
> Here Apelles stopped him (his patience gone):
> Friend, judge things not higher than a boot!

ernment. It seems, however, that his calculations were made without strategic knowledge, and he was forced to withdraw his vanguard from Beirut. Secretary of State Dulles's aggressive-diplomatic 'Eisenhower doctrine' forced Washington directly towards war, bypassing half-war; however, Dulles placed General Norstad in a difficult situation: he had no military units trained and equipped to fight irregular forces (which is why the American intervention force was happy with its inaction, and American strategists were glad to withdraw without significant embarrassment to the US).[12]

Even in a half-war, civilian leaders limit and restrict the decision-making process of strategists. They create a 'War Cabinet' (under Churchill it was overt; under Roosevelt and other dictators it was less explicit and formal), which allows non-military people to take war out of the hands of military specialists. In such a War Cabinet, the voice of strategy sounds very weak, overwhelmed by a choir of ignoramuses who dare to wage a war without any knowledge of military affairs. The best example was Hitler, who overwhelmed the splendid knowledge of German generals with his maniacal mysticism.

Sometimes, it is easier for strategists to triumph over their enemies rather than secure the victory of their strategic concepts over the diplomatic plans of the War Cabinet. Trying to prevent the transfer of the Balkans from the British sphere of influence to that of the Soviets, Churchill insisted on the creation of a 'second front' there, and only after a very difficult struggle was military strategy successful in deploying Eisenhower's forces to Italy via Algeria—from the perspective of strategy this was a more convenient way to victory.

Strategy is forced to fight its way inside the War Cabinet against many astrategic opinions, because war is waged in many theatres: military, diplomatic, political and psychological. Different aims and intentions push through the door of the War Cabinet, forcing strategy away: attempts to secure finance, trade and industry; attempts to enforce support from various neutral countries; attempts to deny similar support from the enemy, thus striking at the opponent with economic and

---

[12] Messner refers here to the 1958 Lebanon Crisis and Operation 'Blue Bat' intended to bolster the pro-Western Lebanese government against internal opposition supported by Syria and Egypt.

STRATEGIYA

political blows; attempts to reinforce the spirit of the nation through various societal actions; and attempts to instil confusion into the souls of enemy soldiers and people.

Strategy likes the shortest and most direct way: final victory is achieved by the destruction of the enemy's armed forces. But how can it march the shortest way if the War Cabinet is advancing circuitously? When Eisenhower was seduced into the conquest of anti-Hitler Bavaria, he did not go to Brandenburg, thus missing the more geo-strategically important prize—Berlin. Hitler reached out towards Ukrainian crops, Kryvyi Rih's ore, Donetsk's coal, and Caucasian oil, removing strategy (as well as all strategists who disagreed with him) from the direct line that Brauchitsch had skilfully drawn directly into the heart of the Red Army.

For centuries, the most important part of geography for strategists was topography and the most important part of ethnography was capital cities, given their role as administrative centres. Nowadays, however, we see the rise of geopolitics with its industrial, military-industrial, political-social and other centres. Since it is very hard to compare the significance of these centres, the formulation of a blueprint for victory has become a very difficult affair. For example, comparing the significance of Moscow and St Petersburg, Napoleon chose the former, thus making his most fatal mistake. How would strategists less talented than Napoleon choose nowadays between dozens of different centres, especially when the psychology of war, like fog, clouds the strategists' vision?

The psychology of strategy is as simple and honest as the psychology of a soldier. The Machiavellian approach to end war by aligning with your enemy, in case you fail to achieve victory alongside your original allies, does not suit strategy's psychology. Hitler, who dodged between anti-Communism, the Molotov–Ribbentrop Pact and Operation Barbarossa, succeeded, because of his hypnotic abilities, in overturning the spiritual base of strategy, but not everyone can be as successful as he was. The military is not an hourglass that needs to be turned in order to act: French forces under the command of General Darlan did not fight Eisenhower very enthusiastically in Algeria in 1942, and the traitorous Marshal Badoglio failed to turn the Italian forces against the Germans.

What would be the psychological base that defined the goals of the Americans, as well as their allies and vassals: a 'liberation war' against the USSR? Would it be a complete conquest or the shattering of Russia? Would it be the Orthodox Church or Eastern Christianity? Would it be a united Yugoslavia or a division between Serbia and Croatia? Would Germany and Poland preserve their old borders or follow the Oder–Neisse line? In this extremely tight Gordian knot, one solution is too complex for a soldier's mind, and another too disappointing for his heart. Therefore, strategy would be forced to bend and take a zigzag course. [...]

The traditional military objective—the enemy's military—has been dispersed into people at war; the traditional geographic objective—the capital—has become one of many surrogate capitals, i.e., different centres of national infrastructure. Vilnius–Smolensk–Moscow was the trajectory of Napoleon's strategic arrow; Moltke's arrow was directed from Sedan to Paris. Today, the strategic idea is perhaps like a ball, kicked by football players from one side of the field to the other: there is one final goal, but there are also many intermediate ones, and they are in different fields: military, diplomatic, economic, socio-political and psychological.

The War Cabinet reduced the role of the supreme strategist, turning him into an executor of the collective will of the cabinet. Moreover, the supreme strategist himself has become a collective. Only his will (within the framework of orders received from the War Cabinet) has remained singular; but his mind, in view of the changing realities, has become collective in nature. [...]

Military headquarters has been swollen not only because of the increasing number of soldiers and units, but also because of the increasing variety of weapons systems: a commander needs aides who are experts in different types of weapons, and each of these experts needs his own headquarter personnel. The supreme commander and the commanders of the various theatres of war (i.e., strategists) cannot perform their duties without a variety of numerous advisers who specialise in non-military elements of strategy: mass psychology, sociological and political problems, economics and diplomacy. While these elements are not under the direction of the strategist, he has to take their influence into consideration. His advisers complement his knowledge and the ability of the all-embracing leader: led by him, they are in fact the brains of the armed forces, the collective strategist who creates military plans.

There are four main types of military plans. The first one is German—massacre—an imitation of Hannibal's 'Cannae', which decisively defeats the enemy on the battlefield (Moltke succeeded in doing so at the Battle of Sedan in 1870; Brauchitsch, in 1941, failed to do so on the Eastern Front, even after a dozen Cannaes).

The second type is English—attrition—baiting the enemy out of reach of his fangs and claws (in the past Marlborough was quite good at this, but Montgomery failed and was forced to engage in a serious fight).

The third type is French—seizing a geographical bulwark (Napoleon I failed with Moscow, but Napoleon III succeeded with Sevastopol).

The last, fourth type, is Anglo-American—terror—the suppression of the enemy spirit, like Attila, who won by frightening. Though Roosevelt failed to frighten the Germans, the collateral 'benefit' of the terrorising bombing of civilians (as well as the destruction of industrial and transport infrastructure) weakened German forces.

While there is nothing new in these four types of military plans, we should anticipate a fifth one—Soviet—the threat of using regular forces with the simultaneous employment of irregular troops and other revolutionary units. [...]

V. <u>Totally Armed People</u>

'Every nation has the army that it deserves.' If we apply this formula of Trotsky's to contemporary nations, then we need to recognise that they do not deserve much.

For centuries, the military had been the expression of the aristocratic spirit, simply because it was shaped and controlled by the aristocracy: royal aristocracy in the time of the Bourbons or democratic during the rule of Napoleon. [...] Then, the military was transformed from aristocratic to parliamentarian [...] unable to infuse people with its spirit anymore, because, in the time of parliaments, the people impose their morality and wisdom onto the military, and not vice versa. [...]

Civilians do incredible things to the military: during the Korean War, the American 'Psychological Operations Command' ordered the payment to every North Korean defector of 5,000 dollars. [...] However, this mercantile spirit of paying for treason did not break the discipline of the Communist dictatorship's military. And it is not surprising, for how would American soldiers react to a similar offer from the Red side?

Nowadays, there is an attempt to replace soldiers' regard for duty with money. For example, American soldiers got used to monetary rewards: during the Second World War, pilots received 'commissions' for every successful bombing; or General Patton fined officers and soldiers for negligence. During the Suez Crisis, the Egyptians offered 100 pounds for the head of a British officer; and the present-day German military has come to the conclusion that fines offer better educative results than any other type of punishment, which damages self-esteem. This is the spirit of our times, and it has to be taken into account.

In the Middle Ages, it was understood that a soldier had to have three main skills: to plunge his sword through an enemy's body, to tame a wild horse, and to conquer a woman's heart. Of these three, however, the only skill that has survived until today is the one that concerns women, because there is no need to tame an engine (it just needs to be tuned), and the skill of handling a sword fills modern soldiers with disgust (a direct killing is not acceptable to the nerves of a modern human being). Even an executioner nowadays is shy to kick a stool away and 'culturally' pushes a lever down to open the hatch under the legs of the guilty victim. Hand-to-hand combat and the bayonet charge were replaced by grenades, not only because they are a better technological solution, but also because soldiers do not have the nerves to kill with their own hands; and with a grenade, it is not me, it is the grenade that kills.

Not only is the recruitment material for the military nowadays of a very low quality [...], but the whole psychological atmosphere in which this military is created is poisonous. For example, in Germany part of the youth protested against the introduction of conscription: 'It is better to be twenty times a traitor than once a dead man.' In England, Toynbee and Russell inspire people with various phrases that can be summarised as: 'Let the Reds occupy us, just do not let them bomb us.' The national instinct of self-preservation, patriotism and sense of civilian duty have been undermined for decades by the propaganda of pacifism and internationalism. Moreover, nowadays these dreamers are reinforced by the protagonists of various Soviet efforts: seventeen different international organisations work under the supervision of the World Peace Council, which established more than 150,000 different committees and councils in seventy-five different states (by 'peace', this council means not only an absolute peace, but also any 'justifiable' expansion of the Soviet Union, which wages only 'just' wars, including an aggressive attack on Finland and a stab in the back of Poland).

The military is not the pride of the nation anymore. It is simply tolerated. To be a soldier used to be the most honourable profession. The inscription on Aeschylus' gravestone makes no mention of his immortal poetry, but commemorates only his military achievements: '... of his noble prowess the grove of Marathon can speak, and the long-haired Persian knows it well'. Nowadays, however, an officer is forced to hide his profession under civilian dress. Thus, it is not surprising that while in 1913 the École Spéciale Militaire de Saint-Cyr had 3,700 applications for 1,000 places, forty years later it had only 400 for 250 places.

The decline in the psychological and physical qualities of human potential results in a reduction in the level of what is asked from a soldier and the military. [...] Within regular forces nowadays, the citizen-soldier is reluctant to subordinate his civil rights to his military duties. While the Communists still dare to train their soldiers to be undemanding and tenacious, the militaries of the democratic states fondle their soldiers with gentle touches and gratify them with comfort, forgetting that Rome fell when its legions had become too soft.

In an attempt to prevent young people from 'wasting' their time in military camps, the length of military service has become ridiculously short. And even though contemporary soldiers should be able to act more independently than ever, there is not enough time to shape their spirit, because most of the time is spent on physical training and education in essential skills. When a German recruit refuses to attend training because he is disgusted by it, the military court decides that the matter is not under its jurisdiction and the civilian court sentences the disobedient citizen to two weeks' imprisonment. When a British soldier runs away from a battlefield, the military court acquits him because the poor fellow failed to control his nerves.

The Nuremberg trials created a new understanding of military discipline: one should follow only those orders which do not conflict with one's conscience. [...] When military officers tried to employ severe methods against the Mau-Mau in Kenya or against even more evil mass killers in Algeria, they were obstructed by their own protesting parliaments and judicial investigators. If a state cannot be severe with an insidious enemy, then it definitely cannot be demanding of its own citizen-soldiers.

Educational methods in contemporary militaries can be very original: in an attempt to make tactical exercises more interesting, a commander may present his soldiers with a beautiful young woman and ask half of them to defend her and the other half to try to abduct her. [...] At the same time, in addition to this and other types of faddishness, militaries try to achieve the best possible results by using psychoanalytic methods, which help to assign every soldier to a position that best suits his temperament, mentality and other psychological characteristics. This approach, however, is quite dangerous, because psychoanalysis can frequently conclude that a soldier is 'unsuitable for the front line', thus justifying simple cowardice and encouraging military judges to acquit very dangerous criminals. [...]

Generally speaking, these days are very idiosyncratic: Marshal Pétain, France's finest soldier, is buried in a prison cemetery, and General de Gaulle, the head of the 'resistance', has become the head of state. This idiosyncrasy shapes the spirit of the army and of its strategy, because the influence of the culture of military values has declined. Moreover, when it comes to strategy, we can see a landslide from the heights of the art of regular war-fighting into the valley of a jazz band of irregular fighting. The totality of people in a country, regardless of their age or gender, participate in a total war, employing every possible weapon: the heavy weapons of the battlefield and the lighter arms of partisan fighting; covert weapons of terror and subversion; and the cheap 'weapon' of the word or the widely accessible 'weapon' of wrecking and sabotage. The employment of traditional weapons is associated with danger and self-sacrifice—something that modern egocentric peoples do not incline to, and strategy therefore gravitates towards a struggle with simpler and more popular types of weapons, thereby changing the face of contemporary war.

VI. A Weapon of Battle and a Weapon of Slaughter

In the time of Peter I, a musket could shoot to a distance of 300 steps and its bullet was accurate at 60 steps. After two centuries of improvement, the problem of distance was solved by the machine gun, and for the short-range battlefield soldiers were given carbines and assault rifles.

The tank, armed with the anti-infantry machine gun and artillery gun, could withstand the pressure of an infantry attack and, inheriting the cavalry's boldness, served as a basis for the creation of striking divisions with an unprecedented degree of manoeuvre. [...] The tank also turned into striking field artillery, though traditional artillery, confident in its power, became a dangerous enemy of the tank—together with mines, bazookas and planes, it robbed the tank of its hegemony.

Artillery finished the last war with the most effective concentration of fire—22,000 guns were employed in the Battle of Berlin—damaging the most distant fortifications. While it has not yet realised the full potential of contemporary gunpowder chemistry and mechanics, artillery already faces a possible demise: after the First World War, the light field artillery gun, the most favourite gun of any artillerist, was replaced by the howitzer; and, nowadays, howitzers have been replaced by missiles launchers, weapons of the lowest quality from the perspective of the art of artillery.

The primitive rockets used by Skobelev to sow fear in the ranks of the Kokandian cavalry[13] have been modernised up to the level of 'Stalin's Organ', and today they can even fly to the moon or circle around the earth, thus becoming the most favoured weapon on land, on sea and in the air.

A thousand years ago, French knights used to say: 'We condemn him who uses a bow as a coward, not being bold enough to attack.' If a bow was considered a weapon that allowed soldiers to avoid contact with the enemy, what can be said about the transcontinental missile?

Putting the moral aspect of this question aside, it is important to state that missiles will remain an amateur weapon of destruction until

---

[13] Messner refers to General Mikhail Skobelev and the use of rocket batteries employed by the Russian Imperial Army in the conquest of the Khanate of Kokand in Central Asia in the late 19th century.

the problem of inaccuracy is solved (this could be quite significant at long distances). Obviously, this 'children's disease' can be solved by increasing the ballistic properties or by radio control [...] and in this way or another, missiles are predicted to have a bright future. It is important to remember, however, that while the spiritual side of military affairs can remain imbalanced, its material side will eventually be settled. Thus, it is just a matter of time until an anti-missile weapon is invented and the missile will meet the fate of all other military-technological 'wunderkinds', becoming just another weapon among all the rest.

Missiles also serve to help guns fight planes in the skies. The aeroplane, however, uses meteoric speeds and stratospheric heights to carry its most powerful weapon, the aerial bomb, to its target. An aerial bomb is significantly more powerful than any artillery shell of similar weight and, undoubtedly, has a significantly larger effective distance. [...]

It seems, however, that contemporary warfare had not limited itself to modernising the weapons of battle, but has introduced new weapons of slaughter. The main difference between these two is that, while the former penetrates (more or less densely) the battlespace with deadly fragments, the latter saturates this space with these fragments, making a human presence impossible. One good example of this shift from the weapons of battle to the weapons of slaughter was the flamethrower, another was napalm. [...]

The military use of chemical gases during the First World War was the first attempt to turn the weapon of battle into a weapon of slaughter. Chemistry provided a variety of different chemical gases suitable for military purposes: irritating, suffocating, both combined (causing a soldier to take his mask off and then poisoning him), extremely painful (affecting the skin and internal organs), overt and covert (initially undetectable, but causing severe illnesses after a set period of time). [...] If militarised chemistry had stayed at the level of 1915, it would have found very wide use even today. Fortunately, however, it has become so frightening that people are afraid of using it. But it is not our humanitarian nature that has led us to avoid the use of chemical weapons (the fanatics of humanism had already destroyed their own ideals in 1918 when they continued the food blockade of hungry Germany): it is our fear of retaliation that restrains our desire to attack.

STRATEGIYA

Another monstrous example of weapons of slaughter is white phosphorus munitions. Thousands of people were burned inside the bomb shelters of Hamburg and Königsberg by the Allies' white phosphorus munitions [...] and those who tried to escape were burned alive like human torches. Previously, the most famous torches were Nero's torches,[14] but now history will remember Churchill's torches (by the way, Sir Winston Churchill was proposed as a candidate for the Nobel Peace Prize).

Thermonuclear weapons are the nonpareil weapons of slaughter. Nonetheless, it was the conventional bombs dropped on Dresden, and not the thermonuclear ones in Hiroshima, that established the record for slaughter, when over one night the air force buried 250,000 women, children and invalids under the ruins of destroyed and burned houses. 'Whoever had forgotten how to cry learned again at the destruction of Dresden,' said the elderly Gerhart Hauptmann. If in a future war someone decides to employ thermonuclear weapons, there will probably be no one left to cry. [...]

People tend to differentiate between strategic and tactical nuclear weapons. While strategic slaughter is considered undesirable, the tactical one is thought to be acceptable—if a sin is not very grave, then it is not a sin. The employment of strategic nuclear weapons will be the peak of the strategy of terror. American strategists calculate how many Americans will be slaughtered by a first Soviet strike, even though these calculations are probably exaggerated to the same degree as the calculations made by the supporters of General Douhet a quarter of a century ago.[15] It is important to remember, however, that people's resilience in a nuclear strike cannot be calculated in advance. While two nuclear bombs and half a million casualties completely destroyed the spirit of Japan, then already losing the war, history offers no evidence to assess people's spirit at the beginning of a nuclear war. It is

---

[14] Messner refers to an 1876 painting by the Polish artist Henryk Siemiradzki, 'Nero's Torches'. It depicts a group of early Christian martyrs who are about to be burned alive as the alleged perpetrators of the Great Fire of Rome, during the reign of Emperor Nero in AD 64.

[15] Messner refers to General Giulio Douhet and his theory of aerial strategy developed in the interwar period.

possible to assume that there are some nations that are capable of withstanding a nuclear strike and there are others that are not, owing to their physical-psychological characteristics. [...]

Neither the invention of gunpowder nor the move from the tactics of bayonet attacks toward a concentration of firepower made such a change in the organisation of forces and tactical ideas as the revolution created by nuclear weapons. Instead of a rigid system of military units and subunits, it created a flexible mechanism for improvising combat groups according to required specifications. Tactical compactness of the forces has been replaced by dispersion, because a compact force will die in a nuclear explosion and there are some chances of survival when a force is dispersed. Instead of the old maxim that getting away from the front offers more security, in the nuclear age the maximum security is to be found at a minimum distance from the enemy, as the enemy cannot deliver a nuclear strike on an adversary in close proximity to its own forces. Instead of sending a division from the front for rest and reinforcement after it loses twenty per cent of its manpower, today, when it is possible to assume that a division can be completely lost in a matter of seconds, the calculations should be not of the losses in divisions, but rather of the losses of divisions as a result of tactical nuclear slaughter.

Optimists, like Fuller, assume that fear of a nuclear bomb will prevent not only its employment, but also the very escalation towards war. Nonetheless, there are known medical cases when a patient committed suicide because he feared his own death.

VII. Military Means

Nowadays, the noble word 'weapon' is associated with a different, truly abominable means of war, such as 'biological weapons' or 'chemical weapons'. [...] In addition, there are such phrases as 'the word as a weapon' or 'psychological weapon'. From the perspective of the publicist, 'the word as a weapon' is a brilliant phrase, but in military jargon 'word' should be described not as a weapon, but as a means. The war cry 'Hooray!' is not a weapon, but a means to give a burst of encouragement to those who bear arms. Psychology is not a weapon either, but merely a means for the better planning of a weapon's employment.

STRATEGIYA

Despite their materialistic nature, even Americans realised the importance of the psyche and during the last war created the Psychological Warfare Division[16] (though it is important to remember that 'psychological war' makes no sense, because there is a big difference between the general definition of 'war' and the specific definition of 'psychological methods to conduct a war'). Such divisions are required not only at the level of the General Staff, but also at the lower levels of the military hierarchy, if not as a full unit, then at least as advisers on the problems related to the fourth dimension of war.

Psycho-strategic plans start, first and foremost, with defining the national aims of war and prioritising them in their order of importance in directing all actions towards the achievement of these aims. All types of government actions should be coordinated with psycho-strategic plans and activities. Secondly, any psycho-strategic plan should strive to uncover the psycho-strategic plan of the adversary and find ways to counteract it. [...]

Military leaders were familiar with military psychology thousands of years before the field of 'military psychology' was created—even barbarous armies employed 'psychological attacks', using their war cries to confuse their enemies, thus increasing the physical power of their attack by psychological means. Another example was the painted dragons on the walls of Chinese cities that were used as a 'psychological defence'. Military psychology teaches that even small events can have enormous psychological consequences: the sinking of the RMS *Lusitania* created such an outcry that it resulted in the United States' declaration of war against Germany. Eight decades ago, during the Franco-Prussian War, the Siege of Paris and the Battle of Gravelotte disturbed neither the Russians, nor the Balkan peoples, nor the Spanish. [...] Nowadays, however, the psychological effect of Rommel's conquest of Tobruk or the capitulation of General Paulus in Stalingrad was felt all across the world, stirring favourable or unfavourable sentiments towards Germany. And these sentiments were followed by real actions, for example, by

---

[16] In fact, the Psychological Warfare Division was a joint Anglo-American organisation set up in the Second World War and tasked with conducting principally 'white' tactical psychological warfare against German troops in northwestern Europe during and after D-Day.

refusing or allowing the loading of coal from a neutral port onto a ship that supplied a German cruiser.

The maze of war psychology has not been illuminated yet by scientific knowledge, and therefore strategists make their way blindly in the dark. They have no choice but to move forward, because in the era of the rule of the masses, the conduct of war has to take the psychology of the masses into consideration.

The aim of the terrorising strategy of the Anglo-Americans was not necessarily of a material nature—the destruction of industrial infrastructure—but more of a psychological one—the destruction of the German people's spirit. This is why only 140,000 bombs were dropped on German industrial targets, and 430,000 on German cities. However, the Anglo-American leadership made the mistake of misjudging the specific qualities of the German people: the suffering of war only reinforced their desire to fight until the end. [...]

It is difficult for psycho-strategists to foresee all possible consequences of their psycho-manoeuvres. The 'untermensch'[17] myth helped inspire Germans to fight against their former friends in the east; when the latter began moving towards Germany in 1944, the population, in panic at the invasion of the 'untermensch', started to flee to the west, disrupting the movement of military forces.

Goebbels should be considered a genius of psychological leadership in times of war. He knew the nature of the German soul. [...] However, Germany failed to understand other nations, and therefore its aims of war led to its defeat in the battle of ideas. 'New Europe' failed in its struggle with 'Against tyranny! For democracy!', and 'the Dismemberment of Russia' was defeated by 'the Great Patriotic War'. Moreover, in the future, anyone who thinks he can split Russia will lose his psycho-battle—the unity of Russia is stronger than the atom.

Not only the main aims of war should be carefully weighed on the scales of psycho-strategy and psycho-tactics, but each of the actions planned or executed too. [...] In the past, psychology at the operational level was not considered by military leaders. The very idea of strategic battles of ideas was incomprehensible. Nowadays, however, the very face of war is shaped by the psychological processes occurring

---

[17] A term used by the Nazis to describe non-Aryan 'inferior people'.

among the masses. Like artillery bombardment preparations, which increase the chances of success of a ground attack, any strategic actions should be preceded by proper psychological preparation.

Today, even tyrants are forced to clothe their high-handedness in a false national enthusiasm. The masses that rule the country (or at least believe that they do) do not like direct orders and, since their leaders do not like to plead, the only solution is to persuade and instil ideas. Roosevelt swore an oath to American mothers that their children would not be sent to fight overseas, but after a short time the same mothers decided that military intervention in Europe was necessary—the idea of an active anti-Nazism was instilled in their minds. Ideas should be hammered like nails—this is the very heart of propaganda. [...]

Self-confident dictators, from Hitler to Nasser, turned the art of propaganda from a secondary strategic means to assist in diplomacy and domestic politics into a great force that not only assisted military interventions, but also made propaganda interventions possible, like those conducted by Nasser in Amman and Damascus.[18]

War in the 20th century is not a clear military affair: it consists of politics no less than tactics. Countries in this war have to be conquered not only by the military, but also by propaganda. Today, nations can deny the fact of physical conquest and continue their spiritual resistance even after the capitulation of the military. Using propaganda, one should pour the elixir of life into one's own masses and poison into the enemy's masses, and by using propaganda as an antidote, one should save one's own masses from the enemy's poison. [...]

In addition to the battle of ideas with an adversary, there is the conflict of domestic political programmes that recalls the old axiom 'the army is outside politics', which actually was the product of a huge misunderstanding. In ancient times, the military (an instrument in a state's politics) was defined as being outside politics, as the word 'politics' was used to define the activities of parties within the state. The military is situated well within politics, but it should be kept separate from parties' activities by the coat of mail of political consciousness to protect it from the political arrows of external as well as domestic enemies.

---

[18] Messner refers to Egyptian propaganda campaigns in Syria and Jordan during the late 1950s.

Offensive or defensive propaganda is doomed to fail if it looks like propaganda. The tone of propaganda should be chosen in accordance with the sensibility and psyche of each nation. Propaganda works for strategy and is guided by psychology.

During the Second World War, in an attempt to damage the morale of American troops, the Germans published propaganda flyers based on a list of American capitalists who were profiteering from the war. Not only did the Yankees not get angry with the capitalists, but since the list included many Jewish surnames, they saw it as another manifestation of Nazi anti-Semitism. In their turn, the Americans, in an attempt to persuade German soldiers to surrender, disseminated flyers that described the comfortable life of German POWs in American camps. German soldiers, however, paid attention not to the comfort of the camps, but to the fact that they were based over the ocean—from which there was no return. When the transfer of POWs to America was replaced by local camps in France, crafty Goebbels, in an attempt to prevent soldiers from surrendering, explained that these camps in France were only transit camps and the final destination was Siberia. Propaganda battles can be very interesting and multifaceted.

The First World War was the first war of propaganda. While the first steps of mass propaganda were unsure, very quickly both sides decided that the first casualty in this war should be the truth and flooded their own countries, as well as those of their adversaries and neutrals, with waves of lies. The propaganda of horrors included Cossacks who ate small children alive or the Sisters of Mercy who mercilessly killed wounded soldiers. When an officer from British intelligence invented a story that the Germans dissolved the bodies of dead enemy soldiers to create fat for candles, lubricants and margarine to feed pigs, it created such a wave of indignation that China declared war on Germany and thousands of American people flooded the recruiting centres.

In the Second World War, however, both sides realised that even a hundred truthful messages would not restore the credibility undermined by one lie. The propaganda of imagined horrors was replaced by the propaganda of threats: Hitler threatened his enemies by hinting that he had a weapon of enormous power; Roosevelt threatened his own citizens by suggesting Germans could bound over the ocean (and

Yankees believed him, despite knowing that Hitler had failed even to bound over the English Channel). People do not believe in something obvious if they do not want to, but they do believe in a 'glorified lie' if they want to.

Propaganda should avoid an open lie, favouring instead a perverse reality and the persuasion of a false perspective. While the BBC was one of the best proponents of this, it was surpassed by the talented German radio host Hans Fritzsche (who was hanged at Nuremberg for his deeds): anticipating British reports about some German failure, he was the first to report it, painting the event in his own colours and then arguing with the interpretation given by the BBC. This is why German people considered him a knight of truth and the BBC the radio of lies.

This style of counteracting propaganda is much more effective than forbidding people to listen to the enemy's radio stations or selling radio receivers that can receive only wavelengths approved by the government (as was the case in the USSR). The struggle of the radio waves has become very fierce, and the energy spent on jamming enemy stations has exceeded the energy spent on radio broadcasting. It is important to remember, however, that technology and propagandists' creativity offer great opportunities if the offensive and defensive propaganda is well organised and directed. Like any military battle that requires a chief military commander, propaganda requires a chief propagandist.

Propaganda by word (radio, speeches or rumours), publications, art, theatre, exhibitions or cinema should be accompanied by propaganda by deed: timely and effective success on the real battlefield significantly affects the psychological battle.

VIII. Military Force

The Great Napoleonic Army consisted of 600,000 soldiers. During the Second World War, the United States mobilised eleven million people and the USSR thirty-three million men and three million women. [...]

The sizes of armies had been steadily growing from one war to another. In 1944 they achieved their maximum size, and have been slowly decreasing since then. In older times, several weapons and ammunition makers could supply a whole army; nowadays, however, a whole 'army' of engineers and workers work hard to supply soldiers with the required military technology and equipment. More and more people 'fight' wars by working far away from the front, and fewer and fewer people actually fight on the battlefields.

Together with the decrease in military manpower, there has been a fundamental change in the traditional organisation of regular forces. Anticipating nuclear strikes, militaries get rid of the administrative rigidness developed in the previous century under the influence of German pedantry when an army consisted of three or four corps, a corps of two or three divisions, a division of three or four brigades, and a brigade of three or four battalions. We can slowly see the disappearance of brigades—a division commander has four or five group-style brigades, among which all battalions and artillery batteries are distributed according to the situation and need. Larger units are assembled not according to the traditional template, but according to circumstances.

The main idea is that each group or unit has to be capable of continuing with its mission independently, in case a nuclear strike eliminates all neighbouring units as well as the higher commands. Military organisation strives to become more and more flexible, creating more independent units and thus increasing their chance of survival. The German principle that 'organisation does not tolerate improvisation' has been archived, because the contemporary battlefield demands improvisation to constantly structure and restructure units. This novelty requires very high intelligence and knowledge from military commanders and common soldiers alike. If in ancient times the army consisted of firm units for hand-to-hand fighting, and in early modern times it consisted of rigid units to create a volley of fire, nowadays the army has become an elastic combination of military machines and a few isolated and independent soldiers. [...]

This novelty changes, reconstructs and rejects ideas that were considered until yesterday as right and necessary. The infantry stops being infantry per se: very few soldiers need to use their feet nowadays, because almost everybody and everything has become mobile with the aid of the combustion engine. The cavalry was removed from the battlefield by machine guns and artillery, and it was removed from the theatre of war by the aeroplane, which indisputably surpassed cavalry in its ability to gather intelligence and raid the enemy's rear.

STRATEGIYA

In addition to its traditional mission of knowing the enemy and its intentions, intelligence units have acquired a new mission: to deprive the enemy of technological surprises and to uncover its secrets, facilitating the invention of counter-technology. The value of intelligence has increased so much that in addition to the marshals of air forces, artillery, etc., we will no doubt see marshals of Intelligence.

Artillery, which ruled the battlefields of 1915–18, has been forced to struggle for its existence, because tanks, missiles and aeroplanes threaten to take its place.

The air force has not been satisfied by its victory over cavalry and its close competition with artillery. Nowadays, the air force, like ground forces, strives to control territory, and for that purpose has created its own infantry—parachute and para-infantry units. Eclipsing the bravest manoeuvres of ground forces by its limitless air strikes, air power comes to the assistance of those who crawl on the ground, using its transport planes to allow massive and rapid operational movements.

There has also been a revolution in naval power. Like the planes that skip over the enemy's front line, submarines skip under enemy radars to control the open sea routes. While it has always been impossible to control the air (excluding the case of unchallenged supremacy of power, similar to one that the Allies enjoyed over Germany in the later stage of the war), nowadays it has become impossible to control the sea (excluding the case of an extreme differential in detection and destruction technologies, similar to the one that occurred in 1944). Instead of controlling the sea by powerful fleets, the navy focuses on escorting convoys of ships and hunting privateers. The significance of sea routes has significantly increased owing to the need to move and transfer men and equipment on an enormous and unprecedented scale. [...]

The revolution in naval strategy has been created not only by the rise of submarines, but also because the 'mosquito' plane killed the 'elephant' dreadnought. [...] Heavy naval artillery batteries are not necessary anymore—naval battles of the future will be waged at distances that significantly exceed any range covered by conventional artillery guns, and the main role will be played by planes taking off from gigantic aircraft carriers. [...] It seems possible that in the future the monopoly of aircraft in sea battles will be overtaken by guided missiles. If this happens, it will revive battles between fleets, bringing the era of aircraft carriers to an end, as they will be too large a target for guided missiles. Such sea duels with guided missiles will be conducted from very long distances, and based on radar identification of targets and missile guiding procedures.

Nowadays, naval commanders have been deprived of the romantic notion of ruling the seas and are burdened by the boredom of convoy activities that tie navies to the coast. On the one hand, transcontinental engagements require solutions to large-scale landing operations, which entrust the navy with very important and worthwhile tasks. On the other hand, the independence of the navy has declined: the navy now serves the army, conquering the coast instead of the sea. The navy has become so obsessed with the coast that previously small corps of boarding troops, intended to board an enemy ship on the open sea, have been transformed into whole divisions of marines, who, after finishing their role as the vanguard of an operation, find themselves fighting together with the army deep inside the enemy's territory.

While the relations between the navy, the air force and the army have been changing, including their cooperation and sizes, these three still constitute the core of the military organisation, i.e., the regular armed forces. [...] Simultaneously, the recently added fourth component of the military—irregular forces—will significantly increase in size and importance on the battlefield. [...]

Irregular forces are divided into two parts: partisan units and subversion groups. The proportion between these two will depend on the geographical and politico-social conditions in every state that wages a war. It will also depend on the character of the nation involved. For example, Germans, who care about an enemy's property as they care for their own, are not suitable for carrying out sabotage; but Poles are natural conspirators and have proved themselves perfect saboteurs; Serbs are traditional partisans, something they have been doing since antiquity; and Arabs are perfect terrorists, who hit you from behind.

During the Second World War, the irregular forces were the product of improvisation, even in the USSR, which pioneered their role in the war. Nonetheless, the scale of irregular war-fighting was enormous. [...]

STRATEGIYA

Irregular warfare is infectious like the Asian flu,[19] and can easily turn into a large-scale phenomenon. It offers common people an easy psychological transition from a normal peacetime political-social struggle into participation in a war-time struggle. Irregular warfare makes possible the total participation of people in a total war.

IX. New Ways of Operatika[20]

Winding its way between different intermediate goals towards its final aim, strategy uses different battles as steps. A battle consists of several subsequent or simultaneous fights, which are directed by operatika, which uses a series of fights as steps towards its goals. A fight consists of hostilities by means of which tactics achieves the goals laid down by operatika.

In the last war the scale of operatika was enormous: fights that led to hundreds of thousands of captives and extensive manoeuvres behind enemy lines were not rare occasions. [...] The significant increase in the fighting power of all military branches made battles and fights very bloody, stressful and prolonged: there was no quick bayonet strike that could solve the problem, and a shooting competition is by its very nature quite a complex and slow affair. [...]

Armed forces have become more mechanised and motorised, leading to a significant increase in the potential scale of military manoeuvres. The manoeuvres of air forces can extend to hundreds of kilometres when supporting land or naval forces; even to thousands of kilometres when performing air raids. The infantry, assisted by trans-

---

[19] Messner refers to an outbreak of avian influenza that originated in China in early 1956 and lasted until 1958. It was called at that time 'Asian Flu'.

[20] 'Operatika' is the original word used by Messner. While the word makes certain sense in Russian and English, implying operational level or operational art, it is not a Russian word. Since Messner did not use it in the sense of 'art' or 'level', it is left here in its original spelling. The absence of the term in Russian military terminology can be explained by the fact that it was introduced by the Russian officers in exile and was never picked up by the Soviet military. In fact, Kersnovski attributes the introduction of this concept to Messner, Anton Kersnovski, *Filosofiya voyny* [The Philosophy of war], (Belgrade: Izdanie 'Tzarskogo Vestnika', 1939), p. 31.

port aviation, can jump over the enemy's head hundreds of kilometres into its rear. The navy can manoeuvre on a grand scale, performing large-scale amphibious operations without, as in the past, having control over the enemy's ports. [...]

The potential scale of operatika if assisted by nuclear weapons is so unprecedentedly vast that it requires a fundamental rethinking and reorganisation of the entire system of relations between defence and offence. The old common rule that a war begins where the previous war ended has lost its relevance. The Soviet Union has not only reorganised its armed forces, it has also divided itself into 'theatres', each one of which controls a group of armies and a territory with industry and people, allowing every 'theatre' to fight in isolation and without communication with the central government and the chief strategists.

The new main principle is that a state should not have a concentration of industry and people should be prepared to evacuate densely populated cities; operatika should be conducted without operational concentrations; tactics without tactical ones. Only dispersion will allow for survival and preserve strategic, operational and tactical capabilities. [...]

Nuclear weapons restrict manoeuvrability. Any manoeuvre consists of swinging a hammer and striking it. The danger of nuclear weapons will compel armed forces to strike not with the hammer or fist, but with fingers spread, so that if some of them are amputated by a nuclear strike, the rest can still function. Everything should be spread across the battlefield: individual soldiers, companies, battalions, artillery, divisions, logistics, supplies, headquarters, and even stores, which are already placed back in the rear. Operational manoeuvre and attack will require that these widely spread forces be gathered into a more compact force to perform a quick attack before the enemy identifies the move and deploys tactical nuclear weapons in the area. Under such conditions, no manoeuvre can be deep or last for too long, because it needs to be fast and performed from the rear. The attack cannot be powerful either, because the ammunition stores are dispersed in the rear and there is neither time nor ability to resupply the manoeuvring forces under the threat of the enemy's nuclear strike. [...]

Such an agglomerative manoeuvre, suddenly put together from previously dispersed military units, will require a flawless performance from the overextended headquarters. [...] Is there a possibility that under such circumstances the armed forces will decide to dig themselves into trenches, as in the First World War? Or will the military adopt the operational concept of the 'sword and shield', where land forces are the 'shield' and the air forces are the 'sword'? Adopting such an operational concept threatens to create a situation of winning fights but losing victory, because victory cannot be achieved if the enemy is not pursued. Even if the air force, serving as a 'sword', can create the conditions for victory, the land forces, the 'shield', would be hard put to achieve victory, because the danger of nuclear weapons deprives soldiers of the burst of energy required to pursue the retreating enemy—instead of motorised manoeuvring into the rear of the enemy, the soldiers will hole up deep in the ground.

The air force, unlike land forces, preserves manoeuvrability, even if it is dispersed across thousands of aerodromes and bases. However, the manoeuvres of air forces independent of land operations have only an indirect impact on operations in destroying the enemy state. Time will tell whether such air operations should focus on destroying transport infrastructure, or return to the terroristic nature of the air raids of the Second World War, purposely targeting the enemy population. Or maybe they should restrict themselves to part-terror operations that target military facilities, inevitably causing massive civilian casualties in near-by locations.

Regardless of the nature of future *operatika*, whether it will revolve around trenches or be based on manoeuvres, it will require high-quality soldiers capable of making the right decisions under conditions of total dispersion. [...] Armed forces continue to improve weapons, and it seems that it is high time to focus on improving the soldiers.

X. Irregular Forces

In 1812, General Kutuzov attacked the communication lines of Napoleon with the 'flying corps' (*corps volant*)[21] of Colonel Alexander

---

[21] The *corps volant* was created in Russia by Peter I in 1701. It was a very light and independent unit consisting of several thousands of cavalrymen and infantrymen assisted by light artillery. It was designed for the interception of communication, actions in the enemy's rear, and pursuit of the enemy.

Figner and General Alexander Seslavin, as well as with smaller units of Cossacks. Although, because of some misunderstanding, history remembers these units as partisans, they were not partisans at all. The real partisans were peasants, who took up arms to fight the enemy. Their early guerrilla actions, as well as the political-psychological characteristics of civil wars in Russia, Spain, Greece and China, the huge variety of irregular conflicts during the period 1939–45, and the bloody conflicts in Indochina, Korea and Malaya—all these inspired military thought to study the phenomenon of struggle not on the battlefield, but among the people. The conservative psychology of military commanders, however, hinders the recognition of the phenomenon discovered by French officers in Vietnam, where tactics, military administration and military psychology were turned upside down. While there is much to study and analyse, the West does not pay enough attention to this phenomenon, and the East cannot fully study and understand it because of its Marxist predisposition.

People stopped being passive spectators and silent victims of armed conflicts, and decided to fight by themselves. The citizen of a free state has got used to the idea of mass opposition to the government. [...] This predisposes him to oppose an occupying power alongside his own military or, equally, to rise against the ruling authority of his country in union with another armed group.

The periphery of these people's participation is occupied by two types of struggle. The first is civilian protest, in which many people of all ages and genders can take part without significant risks. The second is violent protests that involve destruction and sabotage. It involves not only defiance of laws and officials' orders, but also intentional damage to property and, possibly, lives. These two peripheral types of struggle require neither organisation nor external support, because the infectious character of secret political-revolutionary units can inspire a wider public to rise in action. Participation in these two types of struggle can turn into an epidemic, especially when: people have a suitable political-historical mindset; war creates an appropriate psychological atmosphere; and proper psychological-political slogans are taken up by people's minds. The direction of such struggles can be conducted by diplomacy, based on its peacetime experience and in coordination with strategic plans.

STRATEGIYA

While diplomacy is responsible for two peripheral types of struggle, strategy takes control of the next two: subversion and terror. Subversion is the destruction of military infrastructure (stores, communication lines, etc.) as well as non-military infrastructure (electricity stations, oil pipes, etc.). Low-level terror targets soldiers on the streets and roads, the lower echelons of the administration, and people who sympathise with the adversary. High-level terror targets the heads of power, following the principle of the famous Yemelyan Pugachev: 'Hack the columns, the fences will fall by themselves'.[22] Subversion and terror on enemy territory are conducted by specialists deployed to targeted areas by planes or submarines (if simple crossing of borders or the front is impossible). Similar actions are conducted by local subversive and terrorist groups ('fifth columns') or small units sent by partisans or militarised wings of political parties. All these actions depend on the level of tactical or, sometimes, operational organisation.

The fifth type of irregular struggle is partisan warfare or, in other words, military activities conducted by armed units consisting of members of the civilian population. Local units gather together from time to time and only operate at a local level. Permanent units of partisans hide in the mountains or forests. While they have a certain degree of mobility, they do not stray far from their villages, which provide them with supplies, take care of the wounded, and gather intelligence. Externally organised partisan units (i.e., escaped prisoners of war, groups of agents sent on missions, etc.) can be very mobile and can move from one area to another according to the situation or orders received. Sometimes, such units forcibly demand cooperation from the local population, who are usually afraid of supporting strangers. Such coercion, however, does not necessarily create public upset, and sometimes it is even welcomed by the population because forcible mobilisation releases the community from collective responsibility, and a person thus mobilised from individual responsibility.

The sixth and most widespread type of irregular struggle is insurrection, where the fighting is conducted not only by partisan units, but

---

[22] Yemelyan Pugachev, leader of 1773–75 rebellion in the Russian Empire after Catherine II seized power in 1762.

also by the major part of the population, which decides to arm itself and fight against the adversary.

One of the classic examples of the development of irregular struggle through all its types comes from Poland between 1939 and 1945, which involved the Wehrmacht in all stages of struggle up to the sixth, when General Tadeusz Komorowski led 40,000 people in the Warsaw Uprising in the summer of 1944.

It is possible that the development of irregular struggle will be restricted to the initial types only, as it was, for example, in Czechoslovakia during the Second World War, where even subversion did not take root. The development of irregular struggle depends a great deal on national character, the politics of the adversary, and fortuity (e.g., the existence of a talented leadership). [...]

The lessons of the not-so-distant past clearly teach us that in the near future waging war without armed forces, but with partisans, saboteurs, terrorists, subversives, propagandists and other non-soldiers, will significantly increase in scale. Irregular forces are helpless without the support of the traditional military (such as instructors, weapons, medical and logistical support, funding). However, when they receive material and moral support from regular forces, they can turn into a most powerful actor. [...]

Irregular struggle has become a very powerful element of war. Any officer who gives the subject careful analysis will discover a new military-political world, where duty is replaced by fanaticism, courage by cunning, chivalry by cruelty, tradition by improvisation, order by contrariness, hierarchy based on seniority by the promotion of arrivistes, statecraft by opportunistic slogans, inherited ethics by utilitarian calculation, and words of wisdom by the shouts of tumult. This world is still alien to regular officers, even though it has already burst into the world of war. In particular it will reshape strategy: during the last war, irregular struggle developed from tactics to operatika, and now is conquering strategy. It cannot be otherwise when almost the whole French Army is deployed in Algeria to fight hundreds of partisans, thousands of terrorists and millions of Arabs, who conduct sabotage against France.

In some areas of future theatres of war, popular insurrection will create anarchy; in others, sabotage and terrorism will produce turmoil. In some theatres, partisans or popular uprisings will paralyse enemy forces; in others, a combination of all these will turn war into an interminable struggle between political fanatics, insanely killing and destroying each other. In comparison with this, the cold-blooded battles of regular forces will be considered acts of mercy towards an enemy. [...]

This revolution in the nature of war can be stopped only by a counter-revolution. A negative development can be stopped only by a positive one. Neither punitive raids nor the occupation of territory can win against the offensive of an irregular struggle. Only the conquest of people's souls can do so. As Suvorov, when fighting against French revolutionaries, used to say: it is nobility that wins war more than weapons. [...]

XI. The Aims of Logistics

Tactics is the science of using forces. Logistics, however, is the science of the needs of forces. [...]

'In times of war, the muses are silent' is a famous saying from ancient Rome. Nowadays, however, the muses of the arts are recruited for propaganda and the muses of science are mobilised by military logistics. For too long a period Hitler used science according to the old Roman rule, and so he was late with his V-2 rockets, guided missiles and nuclear weapons. The Americans, however, gave their scientists unlimited resources and they very quickly overtook the Germans in technical-military progress. [...]

Logistics takes into consideration all possible needs of the forces, from camouflage paint [...] to the powerful radar station in Samsun that permits observation of activities on the aerodrome in Sevastopol, 500 kilometres away; from a computer brain, which independently calculates the coordinates of enemy planes and broadcasts them to all air defence systems; to guidelines for how soldiers lost in enemy territory without food can feed themselves off the land.

In the last war the catalogue of the American military included millions of items, and consisted of hundreds of volumes. Any trifle could be ordered by using a catalogue number and logistics warehouses would deliver it with the speed and accuracy of a supermarket delivery.

However, the scale of supplies required is increasing progressively, creating much concern among logistics officers. [...] During the Raid on Berlin in 1760, Russian forces used 2.5 tonnes of artillery shells; however, during the Battle of Berlin in 1945, they used 25,000 tonnes. A small jet fighter consumes 2.4 tonnes of fuel per hour, but it takes 20 tonnes per hour to power the eight jet engines of the B-52. In the period between 1943 and 1945, the Red Army annually received from industry 100,000 mortars, 120,000 cannons and 450,000 machine guns. It is assumed that in a Third World War, the USSR will annually produce 40,000 aeroplanes and 35,000 tanks; and the US and the UK will produce 130,000 aeroplanes and 52,000 tanks every year.[23] It seems, however, that even these numbers will not satisfy the requirements of American needs, because during war the demand for military-technical equipment is unlimited.

While the problem of manufacturing should worry national industries, the main problem of the logistic forces is the delivery of 'goods' to the 'clients'. [...] The transportation of equipment has to meet ever-rising demands in ever-worsening conditions, because transportation routes are the most vulnerable links in military geography. Moreover, the nuclear age will even worsen the situation, because it demands the dispersion and decentralisation of everything vulnerable, and warehouses are thus the first to be decentralised. Since, in a war, France would need to properly decentralise its warehouses of weapons, fuel, food and other supplies, it seems that the whole of its territory would not be big enough. Furthermore, these dispersed goods have to be delivered to the forces in a matter of hours, not weeks, as in the past, when military operations were prepared in advance.

An adversary will try to isolate the front from the logistics based in the rear, using artillery shells, aerial bombs and nuclear devices, which would make the transportation of supplies impossible, with or without roads. The only option that would remain for the logistic forces is air transport. The last war saw the first attempts to perform airlifts: the Germans did so in Crete, Stalingrad, Demyansk and Narva; the Americans airlifted supplies from Brazil through Nigeria and Sudan to Egypt. The Soviet blockade of Berlin taught Western air transport to airlift up to 235,000 tonnes of supplies per month. The

---

[23] It is unclear where Messner took these numbers from.

STRATEGIYA

main question is: will air transport be able to airlift the supplies required by future forces? [...]

The successful accomplishment by logistic forces of their mission does not ensure victory. In the last war, even when the supplies were delivered in time, tactics and operational art did not always meet their goals. One needs to recall, though, that sometimes Soviet partisans almost managed to paralyse the German lines of supply. In a future war, irregular struggle and nuclear weapons will be the most dangerous enemies of logistics. The officers of logistic forces will be required to demonstrate not only workable solutions and an eagerness to act, but also real valour and bravery.

XII. A State at War

During a war, armed forces ask industry to create a variety of very complex and sophisticated solutions. For example, the design of a new type of submarine requires 15,000 blueprints. Up to 1945, German weapons designers had developed more than 138 types of guided missiles. It is not surprising that the Americans removed 1,500 tonnes of classified scientific-technical documentation from defeated Germany.

To fulfil the armed forces' requirements, industry not only has to stretch its resources to the maximum, but it also has to neglect the needs of the people. While in 1940 military production formed only fifteen per cent of total German production capacity, in 1944 the war needs absorbed fifty per cent of everything manufactured in Germany. Such capture of industry by military needs shatters the economy of a country. [...]

The Crimean War cost 9 million gold dollars, the Russo-Japanese War 13 million, the First World War 100 billion, and the Second World War 375 billion. In 1945, Germany spent 290 million marks a day.[24] [...]

Economic problems, however, should not be considered the most important ones during a war, regardless of how significant they are. The most important problem is the security of the population and its mobilisation for the war effort. Solving this problem is more important than any economic, political or even strategic setback. [...]

---

[24] It is unclear where Messner took these numbers from.

In the future, every state at war will have to evacuate tens (if not hundreds) of millions of people from areas close to the front. This should be performed in advance or in the first hours of war, and it should apply to everybody living within the proximity of at least a hundred kilometres of the front (to save people from tactical nuclear weapons and the consequences of battles on land and in the air). The same haste should apply to the evacuation of children, the elderly and the infirm from administrative and industrial centres. Such a resettlement of masses of people has to be carefully planned and organised, including food, shelter, medical care and transportation. The flows of relocated people have to be managed in a way that will direct capable workers to the places where they are most required. While controlling these panic-stricken people is a very difficult task, the enemy will try its best to complicate the process even more.

During a war, the population has to be protected by a multifaceted and suitably equipped active air and missile defence system. Moreover, the population has also to be provided, where possible, with passive defence, achieved by dispersing population centres and building shelters. After a bombing, it is necessary to extinguish fires, extract people from ruins, give them medical treatment, feed them and supply them with temporary shelter, and start to rebuild the city and, more importantly, its industry. To do this effectively, it is important to form and prepare local rescue units, as well as special emergency reserves, which should be created by the government and include trains and motorcades of firefighters and rescuers, sappers and medical personnel, food, supplies and appropriate technical equipment.

During a war, the demand for people is enormous. [...] The commander of human resources assigns people to (1) administration, police and other internal security forces; (2) air defence; (3) mobile rescue and emergency reserves; (4) trade, educational, medical, entertainment and other bodies; and (5) productive use in science, various industries and agriculture.

Each peacetime profession is rated according to its usefulness to the war effort. For example, Germany recruited only nine per cent of its miners, but sixty-six per cent of its hairdressers[25]—while it is impos-

---

[25] It is unclear where Messner took these numbers from.

sible to fight without ore, having a frequent haircut is not a wartime necessity. [...]

To save human resources in a future war, children will be raised not by their mothers or nannies, but in nurseries and kindergartens. Family meals will be forbidden, because their preparation simply wastes human productive time, and everyone will eat at public canteens. Small shops will be closed in favour of supermarkets, where a small staff will be able to service the public, offering strictly standardised goods. If the war is waged between democracy and Communism, the latter will have to compromise its tyrannical ideology, as it did in the last war. Democracy will also be forced to compromise its individual freedoms and turn to an almost communistic lifestyle. If democracy wins and its victory defeats ideological Communism, it may be forced to preserve the practical communism developed in wartime for the period of recovery or even longer. Even Oscar Wilde with his 'Prince of Paradox'[26] could not think about a more fascinating paradox than this one!

The dearth of labour will be compensated by the employment of prisoners, the mobilisation of workers from occupied territories, and the hiring of workers from neutral countries [...] as happened in the last war, when vast numbers of Chinese workers were employed in Soviet industry. Nowadays, it will be necessary to resettle millions of workers in order to populate factories, or to relocate factories into areas with a significant surplus of labour.

In the past, the manpower of armed forces was defined first and foremost by strategic requirements, and only secondly by the economic capacity of the state. Nowadays, however, armed forces can recruit from what is left after industry and air defences fill their ranks, and after recruitment to the administrative, internal and emergency forces entrusted with maintaining order, thereby allowing industry to function. Yes, the armed forces are still going to consist of millions, but their numbers will hardly reach the multi-million figure.

In 1944, Hitler recruited fifteen-year-old boys to protect the German skies. In 1945 he replaced them with women, mobilising both into his national militia. [...] Ideological wars are ferocious. Future

---

[26] Messner refers to the novel *The Picture of Dorian Gray* by Oscar Wilde.

wars will be even more ideological and thus even more ferocious. 'Woe to the conquered!' is obsolete and should be replaced by 'Woe to the conquered and the conquerors!' In comparison with contemporary war, the Patriotic War[27] was a failed military saunter of Napoleon across Europe, and the Trojan War was Menelaus' protracted picnic.

Conclusions: On the Ashes of Law

This work required twelve chapters, twelve strokes that sketch the face of contemporary war. This face is more fearful than the fiercest Chinese idols. It scares humanity, but people will have to worship this merciless idol anyway.

Future war will involve a clash between the fanatical calmness of Asia and the energy of African tribes, which are in the process of joining the world, as well as the conscious will of white nations, who abandon their usual frivolity in times of mortal danger. The persistence, organisation and economic power of the Western bloc will conquer the assertiveness of Communism and the stamina of its Russian and Chinese slaves. The only obstacle that the West will struggle to overcome is the destructiveness of nuclear weapons. The densely populated nature of Western nations and their anxiety make them more physically and morally vulnerable in comparison with the USSR.

In starting its chess game with the Soviet Union, Western strategy should never think of a nuclear gambit—it would ultimately lead to checkmate. The threat of a super-powerful nuclear response should induce Western strategists to deter the insane and infamous leaders of the East from initiating strategic nuclear actions. The alliance of freedom and humanitarianism should arm itself for a non-nuclear victory. Such a strategy will allow the West not only to escape destruction, but also to prevent humanity's suicide.

Even without committing a nuclear sin during a future Third World War, humanity will have to implore mercy for committing numerous other sins, while it is still praying for forgiveness for the sins committed during the Second World War.

---

[27] In Russian historiography, the French invasion of Russia in 1812 is called the Patriotic War of 1812. It is not to be confused with the Great Patriotic War of 1941–45.

STRATEGIYA

An international treaty is probably merely a piece of paper. Being a formal agreement, it can, like a commercial one, be terminated. A moral-ethical treaty, however, should never be terminated. Even the 'stiff-necked' Jewish people have always tried to fulfil their covenant with God. In the last war, however, people terminated their covenant with God, bending their moral-ethical obligations. These obligations were formulated and fixed by the 1907 Hague Convention Respecting the Laws and Customs of War on Land. However, already in 1912–14, the experts on international law started to interpret it in different ways: when it comes to saving a state from destruction, these laws and customs can be abandoned; however, in realising tactical and operational goals they have to be strictly observed. But even this 'strict observance' proved itself a recommendation only. For example, immediately after becoming the grand strategist of Britain, Churchill ordered, in complete disregard for article 25 of this convention, the start of the terror bombing of German cities. [...]

While there had been clearly defined laws regarding prisoners of war, they were abandoned in the last war. Prisoners of war were employed in military industries, and thus forced to help their enemy. Moreover, they were brainwashed to turn and fight against their own forces, not to mention the widely accepted practice (by partisans and regular forces) of executing prisoners. [...]

The Hague Convention also laid down rules for people's participation in military activities: partisans were required to wear distinctive badges and were prohibited from taking them off or hiding their weapons; only the legitimate government was allowed to appoint their leaders; partisan units were required to comply with all military laws and customs; after capitulation, all types of resistance were to be halted. None of these were ever followed. Partisans hid their identities; they were frequently led by irresponsible, self-appointed leaders; and they were commanded not by legitimate governments, but by refugee committees (like the Polish one in London) or by representatives of foreign powers (like Churchill's son Randolph, and Brigadier Maclean, who commanded Josip Broz Tito and his partisans); the cunning and cruelty of Italian, French and Polish partisans were simply indescribable; and their resistance did not end with capitulation.

Interestingly, while such 'resistance' considered itself lawful, truly lawful forms of repression intended to eliminate this 'resistance' were declared illegal by American and British judges, who sentenced Germans to death for carrying out their duties. And this despite the fact that the judges knew perfectly well that international conventions did not prohibit such repressive actions. Not to mention that these were, in fact, allowed by American and British military laws as a means to protect their forces from clandestine killers. The same laws considered the execution of hostages the natural answer to the insidious killing of soldiers on the streets.

While the honourable soldier Albert Kesselring was sentenced for alleged crimes committed by his forces in Italy, Charles de Gaulle was not sentenced for the crimes committed by his *maquis*, nor were other ringleaders who raged across Italy, Poland, Yugoslavia and the USSR. [...]

The cruelty towards the civilian population was simply boundless. The first example of this was, unsurprisingly, given by the British, who had always tried to 'improve' war. In 1900, they were the first to establish concentration camps in South Africa, interning hundreds of thousands of Boer women and children, and killing tens of thousands. Hitler established dozens of concentration camps and ordered their operation according to the rules of collective ferocity (against everyone), special ferocity (medical experimentations on humans), and biblical ferocity (racial extermination). [...]

The great institution of international law, founded by Tsar Nicholas II,[28] was destroyed during the First World War. On its ashes, states have been trying (without much inspiration) to build precarious shacks, as they have no desire for anything bigger. Moreover, the Red Cross smuggled into law[29] a concept that does a very poor job of protecting populations, as well as betraying military officers. Without banning civilians from killing soldiers in the streets, the 1949 Convention prohibits the military from taking hostages and conducting repressive actions, thus delivering armed forces into the hands of terrorists. The articles of this convention give anyone who suffered during war an

---

[28] Messner refers to the First Hague Conference in 1899, which was proposed by Tsar Nicholas II, who was instrumental in initiating the conference.

[29] Messner refers to the 1949 Geneva Convention Relative to the Protection of Civilian Persons in Time of War.

STRATEGIYA

option to sue officers for conducting 'wartime crimes', accusing them of taking actions which are the outcome of the essence of war and the nature of the military profession.

The rights of prisoners of war have been restricted again. However, does anyone recognise the right to execute prisoners of war if they continue to fight, despite being captive?

In 1950, the UN developed a 'Collection of Laws that Guarantee Peace and Security'.[30] This masterpiece of hypocrisy defines three groups of crimes: war crimes, crimes against humanity, and crimes against peace. What this masterpiece signifies can be judged by the fact that at Nuremberg, German admirals were accused of building and preparing their fleets for aggression. It seems that today the best recommendation for navies is to build their fleets for fishing and tourism. [...]

How can international law benefit humanity if governments manipulate it as they please? For example, Roosevelt (without declaring war against Germany, though willing to help the British) unilaterally extended American territorial waters from six miles to six hundred.[31] In April 1941, US Admiral Ernest King, having in mind the same idea of helping the British, unilaterally changed the ancient geographical definition of the Western Hemisphere, extending its boundary from 20°W to 26°W.[32] In November 1941 (when the US was not yet at war with Germany), the USS *Omaha* captured a German ship carrying rubber. The American commander justified his piratical actions by referring to an obsolete rule that ordered American ships to capture vessels suspected of involvement in the slave trade.[33]

'Winners are not judged' was intended to mean that those who achieve victory can be forgiven for mistakes conducted during military operations. Nowadays, however, it means that while the winners

---

[30] Messner refers here to the 1950 Principles of International Law Recognized in the Charter of the Nuremberg Tribunal and in the judgment of the Tribunal. It seems that he purposely changed the name to highlight his sarcastic attitude to this document.

[31] Messner probably refers here to the 200–300-nautical-mile-wide neutrality zone declared by Roosevelt in 1939.

[32] Messner refers to Admiral King's Defence Plan no. 3.

[33] Messner refers to *Omaha II* (CL-4).

should not be punished for their unlawful behaviour, the losers should be punished for their lawful actions. Victory or defeat defines the attitude towards 'war crimes'—that is the lesson of this century. [...]

Afterword

Pacifism was a dream of universal virtuousness. It failed when it limited itself to the West without extending its preaching to the aggressive East. Pacifism was a fantasy that betrayed the West when Communism started its campaign.

Conscience, reason and the instinct of preservation are needed to resist Communism, and when other methods fail, resistance by war is inevitable. People will have to give themselves up to the nature of war, regardless of how horrifying its face is. Horror is not horrifying for one who is brave. While humanity can do nothing to lessen the horror of contemporary war, it can avoid the situation where the face of war is reminiscent of the face of Medusa.

The West should reject the strategic as well as tactic planning of war in the nuclear mode. The purpose of nuclear weapons should be a mutual compulsion not to use nuclear weapons. Fear of retaliation restrains one from committing a crime. Even the most dishonourable commander will not send assassins to his enemies, because he fears that the same method of struggle could be used against him.

It is an act of timidity to assume that the West has nothing else but nuclear weapons to secure victory against the Red bloc. The West has many things in its favour that will help it achieve victory against the East: it has a larger population, a better economy and a more developed technical basis. Moreover, it has the most important political advantage: while in the West the poor and anti-national part of the population is only a minority, in the East the destitute and anti-Communist population is the vast majority.

If the threat of total annihilation removed annihilation from the conduct of war, it will limit war by the centuries-old principle of compelling an enemy to stop its resistance by the power of spirit and the threat of force. War will remain ferocious because of its increasing scale, complexity and totality. The heroes of the Trojan War would have been terrified by the style of war conducted by Xerxes I;

STRATEGIYA

Alexander the Great would be lost if he tried to conquer Napoleon; and 'le Petit Caporal' would be surprised by the ability of Paul von Hindenburg, Ferdinand Foch and Dwight D. Eisenhower to conduct operations in vast chaotic wars. It is impossible to control chaos, but it is possible to exercise command in chaotic environments and achieve victory despite the chaos. Bravery is the only option. Before a war it is important not to be afraid of it, and during a war it is important not to lose bravery. The West thinks that Pluto/technology rules the pantheon of the gods of war. Instead, the West needs to believe that the real Zeus among the rest of the gods is valour. Victory is achieved not by weapons, but by brave confidence in your own power. Victory should be believed in. Constantine I won not because he improved his weapons, but because he believed that 'in this sign thou shalt conquer'.

ABOUT THE AUTHORS

General of the Infantry *Genrikh Antonovich Leer* (1829–1904) is considered one of the founding fathers of Russian strategic thinking. At the peak of his career, he was the director of the Nicholas General Staff Academy (1889–98), a member of the Imperial St Petersburg Academy of Sciences, the Royal Swedish Academy of Sciences, and the St Petersburg Imperial University. In 1874, he was one of the senior Russian delegates at the Brussels conference, where he worked with Alexander Jomini (son of Antoine-Henri Jomini) to develop an International Declaration concerning the Laws and Customs of War. During his career, he authored many works on military history, strategy and tactics, some of which were translated into German. Leer died in 1904 in St Petersburg.

Lieutenant General *Evgeny Ivanovich Martynov* (1864–1937) was born in Sveaborg fortress. In 1899, after graduating from the Nicholas General Staff Academy, he served in a series of commanding positions, as well as aide-de-camp at several headquarters within the Russian Imperial Army. During the 1904–5 Russo-Japanese War, Martynov commanded an infantry regiment. For his achievements during this war, he was awarded the Golden Weapon for Bravery and was promoted to major general. In addition to his military career, Martynov was a journalist. In 1913, after he published several articles that exposed bribery and corruption in the army, he was forced to retire from the service. At the beginning of the First World War, Martynov returned to the service. However, as early as 1914, during an aerial reconnaissance mission, he was captured by the Germans. Upon his release in February 1918, he voluntarily joined the Red Army, where he served in different educational and research positions on the General Staff of the Red Army until his retirement in 1928. After the retirement, Martynov continued to contribute to the development of strategic thought by translating foreign strategists and teaching at the Moscow State University. During his career (before and after the 1917 Revolution), he authored numerous works on military history and strategy. On 23 September 1937, Martynov was arrested. Accused of counter-revolutionary propaganda, he was executed a few months later.

General of the Infantry *Nikolai Petrovich Mikhnevich* (1849–1927) started his military career in 1867. During the 1877–78 Russo-Turkish War he was a company commander. From 1882 to 1888 he served in a series of commanding positions, as well as aide-de-camp at several headquarters within the Russian Imperial Army. From 1890 to 1895, Mikhnevich served as chief of staff of the First Cavalry Division. Simultaneously, from 1892, Mikhnevich became a professor at the Nicholas General Staff Academy, presenting lectures not only at the Academy but also at several other military schools and institutions. From 1904 to 1907, promoted to lieutenant general, he was the director of the Nicholas General Staff Academy. On 7 March 1911, Mikhnevich was appointed as a commander of the Grand Staff of the Russian Imperial Army, a position he held until his retirement in April 1917. After the 1917 Revolution, he joined the Red Army, giving lectures at the Mikhailovsky Artillery Academy. He published many works on military history and strategy which influenced a generation of Russian strategists, many of whom continued to serve in the Red Army until Stalin's Great Purge in the late 1930s. He died in Leningrad in 1927.

Anton Antonovich Kersnovski (1907–44) was born in Tzepilovo village, close to the city of Soroca (in present-day Moldova), on a small estate acquired by his grandfather. At the age of thirteen, Kersnovski joined the White Movement. In 1920, he was evacuated from Crimea with the remaining forces of General Wrangel's Russian Army. After a short visit to his village (then in Romania), he moved to Vienna to

ABOUT THE AUTHORS

study at the Diplomatic Academy. After graduating from the Academy, he moved to France where he continued his education, taking occasional courses at the University of Burgundy and the Saint-Cyr Military Academy. Kersnovski's first article, 'On American Artillery', appeared in 1927 in *Russkii Voennyy Vestnik* (Russian Military Herald), published by the Russian émigré community in Belgrade (when its author was twenty years old). His subsequent articles mainly focused on military history and contemporary international relations. At the beginning of the 1930s, he published a series of articles predicting the rise of Hitler and another world war. Since Kersnovski was a private man and did not participate in the social life of the Russian émigrés, no information about him was then available. Commenting on his articles, German experts referred to him as 'russischer General Kersnovski', and some Russian publicists even claimed that Kersnovski did not exist, his name being a pseudonym used by a group of senior Russian generals, as nobody believed that a civilian could write so vividly and knowledgeably on military affairs. With the German invasion of France in 1940, Kersnovski was recruited to the French Army. After he was seriously wounded, he returned to Paris where he died in 1944 from exacerbation of a condition of chronic tuberculosis that dated back to the Civil War. During his short writing career, Kersnovski authored hundreds of articles and numerous books, many of which have been widely republished, analysed and commented on in post-Soviet Russia.

Lieutenant General *Nikolai Nikolayevich Golovin* (1875–1944) came from an old noble family (he was son of General of the Infantry Nikolai Mikhailovich Golovin). After graduating from the Nicholas General Staff Academy in 1900, he served in several military headquarters while continuing his academic career (as professor of the Nicholas General Staff Academy from 1908). Golovin started the First World War as the commander of the Grodnenskii Life Guards Hussar Regiment. He proved himself a talented commander, winning many awards (including the Golden Weapon for Bravery) and the rank of major general. During the Civil War, Golovin joined the White Movement; however, after visiting both Kolchak's Provisional All-Russian Government in Omsk and Wrangel's Russian Army in Crimea, he thought the situation hopeless and emigrated to France.

ABOUT THE AUTHORS

Golovin is considered one of the leaders of the Russian émigrés, as his aristocratic roots not only earned him respect in the Russian community, but also helped to open doors in various European capitals. In the interwar period, with the aim of preserving the scientific traditions of the Russian Imperial Army and raising a new cadre of professional officers, Golovin organised military schools and courses (initially in Paris, then in Belgrade and other main centres of the White émigrés). From 1926 to 1940, Golovin was the official representative of the Hoover Military Library in Paris. In 1930–31, he attended the Military College in Washington and Stanford University in California, where he gave a series of lectures on the history of the First World War. During this time, he published extensively, focusing on military history, strategy and military science. With the occupation of France in 1940, Golovin took up a post in the collaborationist Committee for Mutual Aid of Russian Emigrants. At this point Golovin was engaged in sending Russian volunteers to work in Germany and restaffing the Russian Liberation Army with officers. In 1942–43, he wrote a series of propaganda articles about the victory of the Third Reich. The defeats of Germany, the death of his wife in August 1943, and disappointment with his only son, who joined the British air force as an aeronautical engineer, undermined his health. Golovin died in 1944 of a heart attack.

Colonel *Evgeny Eduardovich Messner* (1891–1974) was born in the Kherson governorate in the southern Ukrainian region of the Russian Empire. His great-grandfather was a Württemberg noble who migrated to Russia during the reign of Catherine the Great. With his father a Lutheran and his mother Catholic, his parents raised him according to the Russian Orthodox tradition. Following his desire to become an officer, Messner, as an external student, passed the final examinations of the Mikhailovsky Artillery Academy in 1912. Starting the First World War as a junior officer, Messner quickly climbed the ladder. In 1916, already a staff captain, Messner was sent to the Nicholas General Staff Academy. In February 1917, Messner went back to the front and was soon appointed to the position of acting chief of staff of a division. Messner joined the White Movement almost immediately after his return to Odessa in the late spring of 1918. He served in various posi-

ABOUT THE AUTHORS

tions during the Civil War, most notably as the last chief of staff of the Kornilov Division of Wrangel's Russian Army. After evacuating from Crimea, Messner moved to Belgrade, where he took an active part in the social, military and academic life of the Russian émigré community. He published many articles and was a lecturer at the School of Higher Military Science Courses, organised by General Golovin in Belgrade. After the start of the Second World War, Messner continued his work in Belgrade, preparing officers for the Russian Liberation Army. Until the spring of 1945, Messner served in the military propaganda department of the Wehrmacht's 'South East Army', where he led the Russian section. In March 1945, Messner took the position of head of the propaganda department in the First Russian National Army, established under the command of Russian émigré General Boris Alexeyevich Smyslovsky-Holmston. After the army's capitulation in Liechtenstein in May 1945, Messner moved to Argentina. After settling in Buenos Aires, Messner continued his earlier work as a journalist, publisher, writer and military theorist. One of his most prominent achievements was the establishment of the South American branch of the Institute for Research into War and Peace in Buenos Aires, named after General Professor Golovin. His publications in Argentina mainly focused on the conceptualisation of war based on his personal experiences and his interpretation of the struggle between the West and Communism (the Soviet Union and China). Messner died in 1974 in Buenos Aires.

INDEX OF NAMES

Achilles, 224
von Aehrenthal, Alois Lexa, 159
Aeschylus, 258
Agamemnon, 224
Aksakov, Sergey, 156
Alekseev, Mikhail, 183, 186, 189
Alexander I of Russia, 152, 185, 236
 and Congress of Vienna, 104
 and France, 82–82
Alexander II the Liberator, 156, 161, 204
Alexander III the Peacemaker, 149
Alexander the Great, XI, 24, 29, 34, 40, 41, 105, 288
Antiochus III the Great, 78
Arakcheyev, Alexey, 185
Archduke Charles, Duke of Teschen, 2, 28, 46, 48, 63
Arthur Wellesley, 1st Duke of Wellington, 170
Attalus III, 78
Attila, 256
Augereau, Charles Pierre, 67
Augustus II the Strong, 27
Augustus, Caesar, 77–8

Badoglio, Pietro, 254
Bain, Alexander, 217
Baratov, Nikolai, 164
Barclay de Tolly, Michael Andreas, 184
Bargation, Pyotr, 156
Bebutov, Vasiliy, 185
Belyaev, Mikhail, 186
von Bennigsen, Levin August, 182
von Berenhorst, Georg Heinrich, 50
Bessières, Albert, 216
Bestuzhev-Ryumin, Alexey, 80
von Bethmann-Hollweg, Theobald, 160
von Bismarck, Otto, 103, 105
 and Austro-Prussian War, 89
 and Ems Dispatch Affair, 141, 159–60
 and press, 95
 and Schnaebele Affair, 149
 and Unification of Germany, 99–100
von Blücher, Geghard Leberecht, 170, 183, 195
Bluntschli, Johann Kaspar, 35
Bogdanovich, Modest, 50

INDEX OF NAMES

Bosquet, Pierre, 171
von Brauchitsch, Walther, 254, 256
Bredov, Nikolai, 167
Briner, Julius, 162
Brusilov, Aleksey, 186
Budyonny, Semyon, 168
von Bülow, Dietrich Heinrich, 2–3, 45–8, 51–2, 63
Byng, Julian, 171

Cadorna, Luigi, 179
Caesar, Gaius Julius, XI, 29, 41, 44, 79, 135
Catherine II the Great, 82–3, 152, 157, 292
  and Poland, 92
  seizure of the throne, 81, 276
  and Turkey, 94, 96
Charles VIII of France, 143
Charles XII of Sweden, 176, 178, 183
Chekov, Pavel, 155–6
Churchill, Randolph, 284
Churchill, Winston, 243, 253
  and bombing, 237, 262, 284
Ciano, Galeazzo, 252
von Clausewitz, Carl, XIII, 1–3, 7, 16, 21, 26, 49, 51, 117, 124, 195, 199–204, 215, 249
  and 'Clausewitzism-Leninism', 141–2, 195
  and German national doctrine, 188, 192–3
Comte, Auguste, 208
Constantine I the Great, 288
Corneille, Pierre, 53
Cortés, Hernán, 244
Courbet, Gustave, 6
Cru, Jean Norton, 204

Danilevich, Adrian, 2, 19

Danilov, Yuri, 174
Darlan, François, 254
von Daun, Leopold Joseph, 184
Davydov, Denis, 156
de Gaulle, Charles, 246–7, 259, 285
de Guibert, Comte, 44
de Marmont, Auguste, 21, 27–8, 68, 70–1
de Montesquieu, Charles-Louis, 44
de Rogniat, Joseph, 47
de Rohan, Henri, 44
de Telleyrand-Périgord, Charles Maurice, 144
de Vaudoncourt, Frédéric François Guillaume, 64
de Villars, Claude Louis Hector, 121
von Decker, Karl, 50
Delbrück, Hans, 2014, 213
Denikin, Anton, 10, 21, 166
Derzhavin, Gavriil, 156
Dokhturov, Dmitry, 188, 194
Dokhturov, Dmitry, 188, 194
Domnin, Igor, 11, 15, 18, 21–22
Dostoyevsky, Feodor, 4, 151, 156, 193
Douhet, Gulio, 262
Dragomirov, Mikhail, 6, 20, 200, 224, 230
du Picq, Ardant, 204, 215–20
Duffy, Christopher, 6, 20
Dulles, John Foster, 253

Eberhardt, Andrei, 164
Einstein, Albert, 237
Eisenhower, Dwight D., 243–4, 249, 254, 288
  Doctrine of, 253

INDEX OF NAMES

Elizabeth of Russia, 80–1, 184
Eugene, Prince of Savoy, XI, 29
Evert, Aleksei, 163, 186

von Falkenhayn, Erich, 154
Ferdinand II, Holy Roman Emperor, 92
Fermor, William, 116, 193
Fichte, Johann Gottlieb, 188, 193
   Addresses to the German Nation, 141, 192
Figner, Alexander, 274–5
FitzRoy Somerset, 1st Baron Raglan, 171
Foch, Ferdinand, 153–4, 180, 195–6, 288
Fouillée, Alfred, 208
Franco, Francisco, 251
von François, Hermann Karl Bruno, 187, 192
Franz Joseph I of Austria, 87
Frederick the Great, XI, 29, 38, 44, 92, 105, 151, 193
   and Seven Year's War, 46, 67, 80, 89
Freedman, Lawrence, XIV, 4, 17–8
French, John, 180
Fritzsche, Hans, 268
Frunze, Mikhail, 2
Fuller, John Frederick Charles, 263

Gallieni, Joseph, 154
Gaxotte, Pierre, 157
Genghis Khan, 60, 238
Gerua, Alexander, 168
Gerua, Boris, 21
Goebbels, Joseph, 249, 265, 267

von Goethe, Johann Wolfgang, 151
Golovin, Nikolai, 9–10, 12, 13, 21–22, 199–233, 235, 291–2
von der Goltz, Colmar Freiherr, 94, 124–5
Gorchakov, Alexander, 186
Gough, Hubert, 171
Grand Duke Nicholas Nikolaevich of Russia the Elder, 98, 186
Grand Duke Nicholas Nikolaevich of Russia the Younger, 154, 174
Grant, Ulysses S., 247
Gripenberg, Oskar Ferdinand, 186
Gurko, Vasily, 164
Gustavus Adolphus of Sweden, XI, 29, 40–1, 60, 92, 115
Gutor, Aleksei, 2

Hannibal, XI, 29, 34, 41, 256
Hauptmann, Gerhart, 262
Hay, Charles, 180, 236
Hegel, Georg Wilhelm Friedrich, 141
Hermogenes, Patriarch of Moscow, 184
Hey, Charles, 180, 236
von Hindenburg, Paul, 153–4, 288
Hitler, Adolf, 176, 238, 249, 253–4, 278, 282, 291
   and concentration camps, 285
   and Operation Barbarossa, 251
   and propaganda, 266–8
Homer, 53, 55
von Hötzendorf, Franz, 159

Ilovaysky, Dmitry, 157

297

INDEX OF NAMES

Itō Hirobumi, 149
Ivan IV the Terrible, 157
Ivanov, Nikolai, 154
Izvolsky, Alexander, 159

James, William, 217
Jaurès, Jean, 146, 238
Joffre, Joseph, 153
John Churchill, 1st Duke of Marlborough, 256
Jomini, Alexander, 289
Jomini, Antoine-Henri, XI, XIII, 2–3, 7, 21, 46, 49, 51–2, 64, 70–1, 289

Kamensky, Nikolay, 187–8
Karl Philipp, Prince of Schwarzenberg, 60, 183
Kashtalinskii, Nikolai, 149
Keller, Fyodor, 186
Kerensky, Alexander, 222
Kersnovski, Anton, XIV, 4, 9, 11–12, 18, 21, 137–197, 272, 290–1
Kesselring, Albert, 285
Khabalov, Sergey, 187
Khmelnytsky, Zynoviy Bohdan, 139
King, Ernest, 286
Kirillov, Vladimir, 238
von Kluck, Alexander, 153–4
Klyuev, Nikolai, 183
Kokoshin, Andrey, 5, 12, 17–20, 22
Kolchak, Alexander, 171–2, 175
Komorowski, Tadeusz, 277
Korff, Nikolai, 219
Kornilov, Lavr, 194, 222, 293
Kornilov, Vladimir, 185
Kuropatkin, Aleksey, 183, 185
  and Lake Naroch Offensive, 163

  and Russo-Japanese War, 156, 162, 186
Kutuzov, Mikhail, 152, 249
  and War of 1812, 83, 104, 185, 274

La Marmora, Alfonso Ferrero, 32
Lauriston, Jacques, 68
Lavisse, Ernest, 157
Lazarev, Ivan, 195
Lechitsky, Platon, 186
Leer, Genrikh, XIII, 2, 4–10, 12, 16, 18–22, 23–73, 204–6, 289
Lenin, Vladimir, XIV, 2, 166, 241
  and 'Clausewitzism-Leninism', 141–2
Leonidas I of Sparta, 247
Liddell Hart, Basil Henry, 5
Lloyd, Henry, 3, 7, 31, 34, 42–4, 49, 51, 53–4, 56–7, 66, 69, 70, 120, 122
Lomonosov, Mikhail, 155–6
Loris-Melikov, Mikhail, 186
Louis Philippe I, King of the French, 144
Louis XII of France, 143
Louis XVI of France, 144
Louis XVIII of France, 84, 144
Lubbock, John, 218
Ludendorff, Erich, 154, 174
  and Clausewitz, 192
  and Great War, 169–72, 175
  and Total War, 237
Lyautey, Hubert, 150

Macaulay, Thomas Babington, 68–9
Macdonald, Étienne, 68, 194
Machiavelli, Niccolò, 163
von Mackensen, August, 154, 183

INDEX OF NAMES

Maclean, Fitzroy, 284
Mangin, Charles Emmanuel Marie, 153–4
Martynov, Evgeny, XIII, 2, 5–10, 19–20, 75–105, 289–90
Massu, Jacques Émile, 247
Maurice, Count of Saxony, 27, 44
Medem, Nikolai, 6, 9, 19, 21, 50
Mendeleev, Dmitri, 151
Menelaus, 283
Menshikov, Alexander, 170
Messner, Evgeny, XIV, 9–10, 13–14, 20–22, 168, 235–288
Miaja, José, 251
Mihailović, Dragoljub-Draža, 241
Mikhalev, Sergey, 2, 17–18.
Mikhnevich, Nikolai, 2, 6–10, 12, 16, 20, 107–136, 205, 290
Mill, John Stuart,
  A System of Logic, 12, 200–3, 206, 209–12
Miloradovich, Mikhail, 188, 194
Milyukov, Pavel, 157
Milyutin, Dmitry, 6, 19, 185, 195, 204
Mithradates VI of Pontus, 78
von Moltke, Helmund, the Elder, 2, 103, 105, 119, 151, 174, 195–6, 255–6
  and Ems Dispatch Affair, 159–60
von Moltke, Helmund, the Younger, 154, 160, 187
Monet, Claude, 6
Montecuccoli, Raimondo, 44
Montgomery, Bernard, 256
von Morgen, Curt, 192–3
Moscow, Advance on, (1919), 165

Mosso, Angelo, 216–7
Mozart, Wolfgang Amadeus, 155
Muravyov-Karsky, Nikolay, 185
Mussolini, Benito, 252

Nakhimov, Pavel, 185
Napoleon I, XI, 16, 20, 26–7, 29–30, 33–4, 36, 38, 43, 50, 53, 61–2, 64–5, 72, 90, 92–3, 105, 113–5, 117, 123, 127, 129, 131–2, 135, 140, 144, 151, 170, 184, 200, 220, 274, 283, 288
  and Alexander I of Russia, 83–4, 96, 104
  and Battle of Fombio, 41
  and Battle of Leipzig, 67–8, 161, 185
  and Campaigns of 1805–9, 32, 48, 52, 60, 87–8, 121, 177
  and Moscow, 254–6
Napoleon III of France, 138, 160, 256
Nasser, Gamal Abdel, 252, 266
Nedić, Milan, 241
Nero, 262
Nevsky, Alexander, 184
Ney, Michel, 27
Nicholas I of Russia, 81, 186
  and Crimean War, 87
  and Hungarian Revolution, 84–5
  and revolution, 84
  and Russian army, 101–2
Nicholas II of Russia, 174, 183
  and The Hague Convention, 236, 285
Nikon, Patriarch of Moscow, 184
Norstad, Lauris, 253

Obolenskii, Dmitry, 216

INDEX OF NAMES

Ogarkov, Nikolai, 2, 4
Orlov, Alexey, 87
Orwell, George, 16, 22

Pasha, Ismail Enver, 164
Pareto, Vilfredo, 212
Paskevich, Ivan, 185
Pasteur, Louis, 151
Patton, George S., 257
Paul I of Russia, 19, 152, 195, 197
  and French Revolution, 81–2
Paulus, Friedrich, 264
Pearce, James, 14–5
Pétain, Philippe, 222, 259
  and capitulation of France, 247
  and Verdun, 164
Peter I the Great, 21, 29, 33, 105, 155, 157, 182, 184, 245, 260, 274
  and initiative, 124, 197
  and Russian army, 189, 195
  and Sweden, 177
  and Turkey, 91
Peter II of Yugoslavia, 249
Peter III of Russia, 80–1
Phidias, 151
Philip V of Macedon, 78
von Phull, Karl Ludwig, 52
Pilsudski, Josef, 166, 168
von Plehwe, Paul, 186
Plutarch, 24
Polybius, 55
Pompey the Great, 78
Poniatowski, Józef, 68
Potemkin, Grigory, 176
Pozharsky, Dmitry, 139
Protopopov, Alexander, 187
Pugachev, Yemelyan, 157, 276
Pushkin, Alexander, 155–6, 193, 252
Putin, Vladimir, 4

Radetsky, Fyodor, 185
von Radetz, Joseph Radetzky, 47
Raynier, Jean, 68
Ribot, Théodule-Armand, 217
Richelieu, Cardinal, 161
Robert Stewart, Viscount Castlereagh, 161
Rocquancourt, Jean-Thomas, 27
Roediger, Alexander, 159
Romeyko-Gurko, Iosif, 185
Rommel, Erwin, 264
von Roon, Albrecht, 105, 159
Roosevelt, Franklin D., 237, 243, 253, 266
  and Germans, 256, 267, 286
Rostopchin, Fyodor, 152
Rousseau, Jean-Jacques, 150, 161
Rumyantsev, Nikolay, 83, 104
Rumyantsev, Pyotr, 156, 182, 184, 193, 195, 197
Rurik, 157
Russell, Bertrand, 257
Rüstow, Friedrich Wilhelm, 50
Ruzsky, Nikolai, 154, 186

Salieri, Antonio, 155
Saltykov, Pyotr, 184
Savinkin, Aleksander, 10, 18, 21–22
Sazonov, Sergei, 165, 178
von Scharnhorst, Gerhard, 187
von Schlichting, Sigismund, 2
von Schlieffen, Alfred, 187, 195
von Seeckt, Hans, 153, 195
Seslavin, Alexander, 275
von Seydlitz, Friedrich Wilhelm, 193

INDEX OF NAMES

Sherman, William Tecumseh, 247
Shumkov, Gerasim, 225
Sierakowski, Zygmunt, 139
Sigismund III of Poland, 138
Sikorski, Władysław, 246
Skobelev, Mikhail, 116, 153, 156, 216, 260
  and Russo-Turkish War, 116, 153, 185, 216
Skopin-Shuisky, Michail, 247
Skrzynecki, Jan Zygmunt, 139
Sokolov, Afanasii, 157
Sokolovsky, Vasily, XIII, 2, 17, 19
Soult, Jean-de-Dieu, 68
Spencer, Herbert, 227
Stalin, Joseph, XII, 8, 238
Stronin, Aleksander, 123
Stuart, Charles Edward, 31
Sukhomlinov, Vladimir, 150
Sulla Felix, Lucius Cornelius, 78
Suvorov, Alexander, 9, 29, 31, 65, 116, 125, 129, 132–3, 135, 151, 156, 176, 182, 188, 192, 196–7, 278
  and Battle of Trebbia, 180
  and Campaign in Italy, 113
  Science of Victory, 155, 194–5, 214
  and the Alps, 177, 184
Svechin, Alexander, 2, 4, 8, 17–19

Tamerlane, 150
Ter-Gukasov, Arshak, 186
Thiers, Adolphe, 103
Tito, Josip Broz, 241, 246, 249, 284
Tolbukhin, Fedor, 243

Tolstoy, Leo, 200
Tolstoy, Pyotr, 91
Toynbee, Arnold J., 257
von Treitschke, Heinrich, 192
  and German national doctrine, 193, 188
Trotsky, Leon, 247, 256
Truman, Harry S., 244
Tsitsianov, Pavel, 195
Turenne (Viscount of), Henri de La Tour d'Auvergne, XI, 29, 44, 123
Tylor, Edward Burnett, 218

Vannovsky, Pyotr, 185
von Verdy du Vernois, Julius, 196
Victor-Perrin, Claude, 67
Viviani, René, 150
Voroshilov, Kliment, 246
Voyde, Carl, 200
Vvedensky, Alexander, 210

Wagner, Reinhold, 50
von Weyrother, Franz, 52
Wilde, Oscar, 282
William I of Prussia, 100, 105, 159, 160
von Willisen, Karl Wilhelm, 49
Wilson, Woodrow, 240
Witte, Sergey, 162
Wrangel, Pyotr, 166, 168, 175
  in Crimea, 166
  and Civil War, 169, 171–2

Xenophon, 55
Xerxes I the Great, 287
von Xylander, Joseph, 48, 63

Yazykov, Peter, 2

INDEX OF NAMES

Yeltsin, Boris, 11
Yermolov, Aleksey, 150
Yudenich, Nikolai, 153, 186, 194

Zeus, 151, 288
Zhilinsky, Yakov, 154, 223
Zhukov, Georgy, 2
Zhukovsky, Vasily, 156

INDEX OF PLACES, WARS AND BATTLES

Afghanistan, 14
Africa, 238, 244, 249, 285
Åland, Congress of, (1718–19), 178
Alesia, Battle of, (52 BC), 25
Alexandria, 78
Algeria, 246–8, 253–4, 259, 277
Algerian War, (1954–62), 247
Algiers, Battle of, (1956–7), 247
Alps, The
  and Alexander Suvorov, 177, 184, 194
Alsace, 80
America, 244, 267
Amman, 266
Amsterdam, Convention of, (1717), 178
Anatolia, 40
Andorra, 248
Antwerp, 90
Arcole, Battle of, (1796), 134
Armenia, 78
Arno, River, 41
Asia, 24, 108, 238, 244, 283
  Central, 108, 120, 260
  Minor, 78, 186
Aspern-Essling, Battle of, (1809), 38

Austerlitz, Battle of, (1805), 32, 58, 121, 185, 236
Austria, 24, 46, 80, 83–5, 87–90, 96, 98–100, 104, 243
  see also Austro-Hungary; Austrian Empire
Austro-Hungary, 159, 172–4
  see also Austria, Austrian Empire
Austro-Prussian War, (1866), 24, 96, 109, 159
  and national unification, 80, 108

Balkan Peninsula, 78, 91
  see also Balkans
  and Russia, 98–99
Balkan War, First, (1812–3), 108
Balkans, 97–8, 243, 249, 253
  see also Balkan Peninsula
Baltic Sea, 104
Barbarossa, Operation, (1941), 251, 254
Bavaria, 96, 254
Beirut, 248, 253
Belgium, 84, 90, 104
Belgrade, 18, 137, 214, 247, 272, 291–3
Berlin, 21, 45, 49, 50, 82, 100, 159–60, 162, 204, 249, 254, 279

303

INDEX OF PLACES, WARS AND BATTLES

Battle of, (1945), 260, 279
Congress of, (1878), 98
Raid on, (1760), 278–9
Black Sea Straits, 96
  see also Bosporus Straits;
  Dardanelles Strait
Black Sea, 87, 99
  and Catherine II the Great, 81, 94, 96
  Fleet, 164
Bolivia, 243
Borodino, Battle of, (1812), 134, 147, 236
Bosnia and Herzegovina, 159
Bosporus Straits, 97–9, 174
  see also Black Sea Straits;
  Dardanelles Strait
Brandenburg, 254
Brazil, 279
Brest-Litovsk, Treaty of (1918), 140
Brienne, Battle of, (1814), 183
Britain, 170, 173, 243, 248, 251–2, 284
  see also England; UK
Brusilov Offensive, (1915), 186
Brussels, 49, 214
  Conference of, (1874), 289
Buenos Aires, 13, 22, 235, 293
Bulgaria, 96–8, 174
Burma, 238

Cannae, Battle of, (216 BC), 41, 256
Carpathian, Mountains, 85, 104
Carthage, 77–78
Caucasian War, (1817–1864), 150
Caucasus, 150, 153, 195
  Campaign, (1914–8), 153
Cetinje, 247

Champaubert, 47
Channel, English, 238, 268
Chechnya, 14
Chemin des Dames, 170–1, 216
China, 4, 161, 243, 248, 267, 272, 275, 293
Civil War,
  American, (1861–65), 108–9, 238, 247
  Russian, (1917–22), 10–1, 13, 138, 153, 165, 168–9, 171, 183, 194, 291, 293
  Spanish, (1936–39), 251
Cold War, (1945–91), XII, 4, 13, 14
Constantinople, 174–5, 243
  and Russo-Turkish War, 97–8, 110, 161, 176
Courland, 174
Cremona, Battle of, (1702), 38
Crete, 279
Crimea, 90, 101, 166–8, 170–1, 177, 186, 290–1, 293
Crimean War, (1853–6), 85, 87, 90, 93, 108, 165, 167–8, 170–1, 177, 181, 185, 280
Custoza, Battle of, (1866), 32
Cyprus, 246, 248
Czechoslovakia, 277

Damascus, 266
Danube, River, 77, 87, 98, 185–6
Dardanelles Strait, 98–9, 164, 243
  see also Black Sea Straits;
  Bosporus Straits
Demyansk, 279
Denmark, 80–1, 89, 100, 177–8, 241
Dinant, Battle of, (1914), 181
Donetsk, 254

304

INDEX OF PLACES, WARS AND BATTLES

Dresden, 238, 262
  Battle of, (1813), 58

East Prussian Campaign, (1914), 154
Egypt, 78, 242, 248, 253, 279
Elba, 84
Empire,
  Austrian, 84, *see also* Austria; Austro-Hungary
  German, 80, 188, *see also* Germany
  Holy Roman, 27, 243
  of Japan, 162, *see also* Japan
  Ottoman, 95, 164 *see also* Turkey
England, 30, 68, 69, 83–4, 87, 90, 99, 177–8, 257
  see also Britain; UK
Eurasia, 238
Europe, 3, 13, 31, 42–4, 77, 81–3, 85, 96–7, 99, 101–2, 104, 143, 146, 161, 166, 195, 238, 242–4, 266, 283
  Central, 166
  New, 265
  Northwestern, 264
  Western, 116
Eylau, Battle of, (1807), 176

Finland, 258
Finnish War, (1808–09), 187
Fombio, Battle of, (1796), 41
Fontenoy, Battle of, (1745), 180, 236
Formosa Strait, 248
Fort Bard, Siege of, (1800), 38
France, 19, 27, 30, 82–4, 89–90, 92, 103–4, 114, 120, 143, 146, 149–50, 159, 161, 166–7, 172, 178, 213, 243, 246–8, 250–2, 259, 267, 277, 279, 291–2

Franco-Austrian War, (1859), 180
Franco-Prussian War, (1870–1), 80, 89–90, 96, 111, 114, 120, 159, 177, 181, 196, 203, 264
Friedland, Battle of, (1807), 32

Galicia, 85, 104, 174
  Battle of (1914), 186
Gallic Wars, (58–51 BC), 79
Gangut, Battle of, (1714), 140
Gaul, 79
Geneva, 44, 148
  Convention, (1949), 285
Georgia, 195
Germany, 42, 79–80, 84, 88, 90, 92, 95, 99, 103–4, 140, 149–50, 159–60, 173, 248, 251, 255, 257, 261, 264–5, 267, 270, 280–1, 286, 292
  see also Prussia
Golden Horn, 164
Gorlice–Tarnów Offensive, (1915), 140, 174
Granicus, Battle of, (334 BC), 40
Gravelotte, Battle of, (1870), 147, 264
Great Northern War, (1700–21), 178
Great Patriotic War, (1941–1945), 265, 283
Great War, (1914–18), 13, 141, 143, 147, 149, 153, 162–4, 170, 172–4, 179, 186, 189
  see also World War, First
Greece, 40, 77, 92, 96, 174, 248, 275
Grunwald, Battle of, (1410), 189

Hamburg, 45, 262

INDEX OF PLACES, WARS AND BATTLES

Hanover, 80, 89, 96
Hesse-Cassel, 80, 96
Hesse-Darmstadt, 96
Hesse-Nassau, 80
Hesse, 89
Hiroshima, 237, 239–40, 262
Holland, 42, 241
  see also Netherlands
Holstein, 79, 81
Hungary, 84–5, 96, 138, 159, 168, 241

Ice, Battle on, (1242), 140, 189
Illyria, 78
Indochina, 275
  War, First, (1946–58), 247
Indonesia, 242
Inkerman, Battle of, (1854), 170–1
Isonzo, Tenth Battle of, (1917), 179
Israel, 248
Italian Campaign, (1796), 177
Italian War of Independence, Second, (1859), 108
Italy, 47, 77, 79, 82, 84, 92, 96, 113, 121, 143, 146, 163, 177–9, 241–3, 249, 251–3, 285
Izmail, Siege of, (1790), 184

January Uprising, (1863–4), 139
Japan, 149, 161–2, 262
  see also Empire, of Japan.
Jena-Auerstedt, Battle of, (1806), 140
Jihlava, 32
Jordan, 266
Jugurthine War, (112–105 BC), 78

Kagul, Battle of, (1770), 194

Kars, Siege of, (1855), 185
Katzbach, Battle of, (1813), 161
Kenya, 259
Khmelnytsky Uprising, (1648–54), 139
Koblenz, 103
Kolberg, Siege of, (1759–61), 184
Kolín, Battle of, (1757), 38
Königgrätz, Battle of, (1866), 58
Königsberg, 262
Korea, 161–2, 248, 251, 275
  North, 243
  South, 243
Korean War, (1950–53), 256
Kryvyi Rih, 254
Kuban Campaigns, (1918), 165, 167
Küçük Kaynarca, Treaty of, (1774), 95
Kurekdere, Battle of, (1854), 185

Lebanon, 242, 248, 252
  Crisis, (1958), 253
Leipzig, 67–9
  Battle of, (1813), 67, 161, 177, 185
Leuthen, Battle of, (1757), 46
Leuven, Battle of, (1914), 181
Liaoyang, Battle of, (1904), 147
Libya, 242
Ligny, Battle of, (1815), 170
Lindenau, 67
Lithuania, 174
Livonian War, (1558–83), 140
Ljubljana, 247
Łódz, 186
London, 13, 17–20, 42, 82, 94, 162, 200–1, 216, 217–8, 241, 284

INDEX OF PLACES, WARS AND BATTLES

Treaty of, (1915), 162, 178
Lorraine, 80, 103
Lützen, 67–8
Lützen, Battle of, (1813), 83

Macedonia, 78
Magenta, Battle of, (1859), 58
Magnesia, Battle of, (189 BC), 78
Mainz, 103
Malaya, 275
Manchuria, 149
Marathon, 258
Marathon, Battle of, (490 BC), 25
Marmara, Sea of, 98
Massachusetts, 204
Mediterranean Sea, 250, 252
and politics of Rome, 77–9
Metz, 103
Mexico, 138
Michael, Operation, (1918), 170–1
Middle East, 250
Milan, 47
Mississippi, River, 3
Moldavia, 87
Molotov–Ribbentrop Pact, (1939), 254
Monte Cassino, 238
Montmitail, Battle of, (1814), 183
Morocco, 150
Moscow, 17–22, 121, 139, 152, 162, 184–5, 208, 218, 254–6, 290
Moselle, River, 103

Nagasaki, 237
Napoleonic Wars, (1803–15), XI, 27, 47–8, 52, 152, 156, 185, 188

Narva, 279
Naroch, Lake, Offensive, (1916), 163–5, 178
Netherlands, 92, 146
see also Holland
Neva, River, Battle of, (1240), 140, 189
New York, 18, 212, 217
and United Nations, 242
Nigeria, 279
Northern Taurida Operation, (1920), 165, 167
November Uprising, (1830–1), 84, 139
Novgorod, 140
Novorossiysk, 164, 166–7
Nuremberg, 141, 160, 259, 268, 286

Ochakov, 176
Odessa, 98, 164, 174, 292
Oravais, Battle of, (1808), 187
Ourcq, River, 153

Paraguay, 243
Paris, 21–2, 28, 46, 103, 153, 162, 167, 199, 204, 216, 230, 247, 255, 291–2
Pact of, (1928), 194
Siege of, (1870–1), 264
Treaty of, (1856), 96
Patriotic War, (1812), 188, 194, 283
Peenemünde, 40
Peninsular War, (1808–14), 27, 47
Pergamon, 78
Persia, 116, 164
Petrograd, 187, 210
see also Saint Petersburg
Piacenza, 41, 43
Picardy, 170–1

INDEX OF PLACES, WARS AND BATTLES

Pläswitz, Truce of, (1813), 41
Poland, 81, 84, 92, 138–40, 165–8, 177, 246–7, 255, 258, 277, 285
Polish Succession, War of (1733–38), 121
Polish–Muscovite War, (1605–18), 139
Polotsk, 47
Poltava, Battle of, (1709), 140
Pomerania, 80
Pontus, 78
Port Arthur, 149, 162
Port Said, 248
Portugal, 243
Prague, Battle of, (1757), 68
Pressburg, Treaty of, (1806), 88–89
Prussia, 24, 42, 52, 79–80, 83–5, 87–90, 96, 99–100, 104–5, 111, 120, 140, 160, 162–3, 172–3, 177–8, 189, 223
see also Germany
Prut, River, 104
Pugachev Rebellion, (1773–5), 157, 276
Punic Wars, (264–146 BC), 77–78

Regensburg, 32
Revolution,
French, (1789–99), 27, 47, 52, 81–2, 93, 144, 152, 181, 236, 241–2, 245
Hungarian, (1848), 96
July, (1830), 84
Russian (Bolshevik), (1917), XII, 8, 10, 12, 143, 222, 290
Rhine, River, 3, 48, 68, 103, 121, 146
Riga, Peace of, (1921), 140, 168
Romania, 98, 241, 290

Rome, 76–9, 81, 92, 116, 249, 258, 262, 278
Russo-Japanese War, (1904–5), 149, 156, 159, 162, 186, 225, 280, 289
Russo-Turkish War, (1877–8), 110, 116, 153, 176, 184–6, 194, 216
Ruthenia, 138
Rymnik, Battle of, (1789), 194

Saint Petersburg, 19–20, 23, 49–50, 75, 98, 107, 162, 184, 204–5, 215, 217, 219, 225, 254, 289
see also Petrograd
Treaty of, (1762), 162
Treaty of, (1875), 162
Treaty of, (1881), 162
Samsun, 278
San River, 104, 174
Santa Lucia, Battle of, (1848), 47
Sardinia, 90
Saxony, 27, 89
Schleswig, 79
Second Schleswig War, (1864), 79, 159
Sedan, 255
Battle of, (1870), 247, 256
Seine, River, 3
Serbia, 96, 241, 255
Sevastopol, 164, 174, 185, 256, 278
Siege of, (1854–5), 185
Seven Years' War, (1756–63), 80–1, 89, 111, 116, 140, 162, 184, 193
Shipka Pass, Battle of, (1877–78), 185
Shumen, 98
Siberia, 171, 267

INDEX OF PLACES, WARS AND BATTLES

Sinai Peninsula, 248
Skopje, 247
Smolensk, 255
Soissons, Battle of, (1814), 38
Soviet Union, 242, 248, 258, 273, 283, 293
  see also USSR
Spain, 27, 78, 92, 159, 170, 243, 275
Spanish Succession, War of (1701–14), 25
Stalingrad, 171, 264, 279
  see also Tsaritsyn; Volgograd
Stallupönen, Battle of, (1914), 187
Stołowicze, Battle of, (1771), 177
Strasbourg, 103
Sudan, 279
Suez Crisis, (1956–7), 247, 257
Sweden, 40, 176–8, 183
Switzerland, 90, 92, 104, 146
Syria, 253, 266

Taranto, 77
Thames, River, 3
The Hague, 149
  Convention, (1899), 105, 236, 285
  Convention, (1907), 284
Theodosia, 164
Thermopylae, 247
Thirty Years' War, (1618–48), 92
Ticino, River, 47
Tilsit, Treaties of, (1807), 24, 83, 88, 89, 140
Tobruk, 264
Trebbia,
  Battle of, (1799), 134, 180, 194
  River, 194
Trojan War, 283, 287

Tsaritsyn, 171–2
  see also Stalingrad; Volgograd
Turkey, 87, 90–1, 93–7, 183, 161, 164, 173, 243, 248–9
  see also Empire, Ottoman

Ufa, 172
UK, 279
  see also, Britain; England
Ukraine, 18, 167
Ulm Campaign, (1805), 32, 48
United Arab Republic, 248
United States, 147, 242, 264, 268
  see also US
US, 240, 250, 253, 279, 286
  see also United States
USSR, 251, 255, 268, 271, 279, 283, 285
  see also Soviet Union

Valenza, 41
Varna, 98
Verdun, 147, 163–4, 178
Verona, 47
Versailles, Treaty of, (1919), 140
Vienna, 32, 48, 82, 87, 113, 121, 162, 290
  Congress of, (1814–15), 84–5, 104, 161
Vietnam, 247–8, 275
Vilnius, 255
Vistula, River, 104
Vitebsk, 47
Vladivostok, 162
Volga, River, 3, 171
Volgograd, 171
  see also Stalingrad; Tsaritsyn
Vosges, Mountains, 103
Vyškov, 32

Wagram, 121, 220

INDEX OF PLACES, WARS AND BATTLES

Wallachia, 87
Warsaw, 84, 104, 139–40, 167
  Uprising, (1944), 177
Washington, 20, 242, 248, 253, 292
Waterloo, Battle of, (1815), 170
Wesenberg, Battle of, (1268), 189
World War, First, (1914–18), XII-XIII, 10, 12, 21, 137, 150, 152, 153–4, 164, 166, 168, 174, 178, 180, 183, 186–7, 192, 194, 203, 223, 229, 240, 260–1, 267, 274, 280, 285, 289, 291–2
  see also Great War

World War, Second, (1939–45), XIV, 13, 238, 243, 246–7, 257, 264, 267–8, 271, 274, 277, 280, 283, 293
World War, Third, 279, 283
Württemberg, 96, 292

Yalu, River, Battle of, (1904), 149, 161
Ypres, First Battle of, (1914), 180
Yugoslavia, 247, 249, 255, 285

Zagreb, 247
Zorndorf, Battle of, (1758), 116, 193